Albert Hassall, Charles Wardell Stiles

A Revision of the Adult Cestodes of Cattle, Sheep and Allied Animals

Albert Hassall, Charles Wardell Stiles

A Revision of the Adult Cestodes of Cattle, Sheep and Allied Animals

ISBN/EAN: 9783337243135

Printed in Europe, USA, Canada, Australia, Japan

Cover: Foto ©berggeist007 / pixelio.de

More available books at **www.hansebooks.com**

U.S. DEPARTMENT OF AGRICULTURE.

BUREAU OF ANIMAL INDUSTRY.

BULLETIN No. 4.

A REVISION OF THE ADULT CESTODES

OF

CATTLE, SHEEP, AND ALLIED ANIMALS.

PREPARED UNDER THE DIRECTION OF DR. D. E. SALMON, CHIEF OF THE BUREAU OF ANIMAL INDUSTRY

BY

C. W. STILES, PH. D., AND ALBERT HASSALL, M. R. C. V. S.

PUBLISHED BY AUTHORITY OF THE SECRETARY OF AGRICULTURE.

WASHINGTON:
GOVERNMENT PRINTING OFFICE.
1893.

CONTENTS.

	Page.
Letter of transmittal	9
Letter of submittal	11
PART I. MONIEZIA R. Blanchard, 1891	15–54
A. Planissima group (*M. planissima, M. Benedeni,* and *M. Neumanni*)	15–26
(1) *Moniezia planissima* Stiles and Hassall, 1892	15–22
Synonymy	15
Hosts	15
Geographical distribution	15
Literature	15
Historical review	15
Anatomy	16
Specific diagnosis	22
(2) *Moniezia Benedeni* (Moniez, 1879) R. Bl., 1891	22–25
Synonymy	22
Hosts	22
Geographical distribution	22
Literature	23
Historical review	23
Anatomy	24
Specific diagnosis	25
(3) *Moniezia Neumanni* Moniez, 1891	25–26
Host	25
Geographical distribution	25
Literature	25
Historical review	25
Anatomy	25
Specific diagnosis	26
B. Expansa group (*M. expansa, M. oblongiceps,* and *M. trigonophora*)	26–42
(4) *Moniezia expansa* (R., 1810) R. Bl., 1891	26–34
Synonymy	26
Hosts	26
Geographical distribution	27
Literature (1810–1892)	27
Historical review	28
Anatomy	31
Specific diagnosis	34
(5) *Moniezia oblongiceps* sp. n., 1893	35–36
Host	35
Geographical distribution	35
Anatomy	35
Specific diagnosis	36
(6) *Moniezia trigonophora* sp. n., 1893	37–42
Synonymy	37

PART I. MONIEZIA R. Blanchard, 1891—Continued.

  B. Expansa group—Continued.   **Page.**

      Host... 37

      Geographical distribution................................. 37

      Literature.. 37

      Historical review 37

      Anatomy.. 37

      Specific diagnosis... 42

  C. Denticulata group (*M. denticulata* and *M. alba*) 42–51

    (7) *Moniezia denticulata* (R., 1810) R. Bl., 1891...................... 42–47

      Synonymy 42

      Host.. 42

      Geographical distribution................................. 42

      Literature.. 42

      Historical review 43

      Specific diagnosis.., ... 46

    (8) *Moniezia alba* (Perroncito, 1879) R. Bl., 1891 47–51

      Synonymy 47

      Hosts... 47

      Geographical distribution................................. 47

      Literature.. 47

      Historical review 47

      Anatomy.. 49

      Specific diagnosis... 51

  D. *Moniezia.* (Undetermined specimens)................................. 51

    General summary of the genus *Moniezia*......................... 51–54

      Generic diagnosis.. 54

PART II. THYSANOSOMA Dies., 1834. 55

    (9) *Thysanosoma actinioides* Dies., 1831 55–59

      Synonymy 55

      Hosts... 55

      Geographical distribution................................. 55

      Literature.. 55

      Historical review... 55

      Anatomy 57

      Specific diagnosis... 58

    (10) *Thysanosoma Giardi* (Riv., 1878) Stiles, 1893...................... 59–69

      Synonymy 59

      Hosts... 59

      Geographical distribution................................. 59

      Literature.. 59

      Historical review 60

      Anatomy 62

      Specific diagnosis... 69

      The systematic position of *Tænia Giardi*.................. 69

      General consideration in regard to *Thysanosoma Giardi* and *Th.*

        actinioides 70

          Generic diagnosis...................................... 70

    (11) *Tænia marmotæ* Froelich, 1802 71–72

PART III. STILESIA RAILLIET, 1893 MS.. 73–82

    (12) *Stilesia globipunctata* (Riv., 1874) Railliet, 1893 MS............... 73–79

      Synonymy 73

      Hosts... 73

      Geographical distribution................................. 73

      Literature.. 73

PART III. STILESIA Railliet, 1893—Continued.   Page.
        Historical review 73
        Anatomy 74
        Specific diagnosis.. 79
  (13) *Stilesia centripunctata* (Riv., 1874) Railliet, 1893 MS 79–81
        Synonymy 79
        Hosts.. 79
        Geographical distribution 79
        Literature 79
        Historical review 80
        Anatomy... 80
        Specific diagnosis... 81
        General remarks in regard to *Stilesia globipunctata* and *Stilesia*
          centripunctata 82
            Generic diagnosis 82
Part IV.   *Species inquirendæ*... 83
  (14) *Moniezia nullicollis* Moniez, 1891................................. 83
        Host... 83
        Geographical distribution 83
        Literature 83
        Historical review 83
        Diagnosis.. 83
  (15) *Tænia Vogti* Moniez, 1879....................................... 83
        Synonymy 83
        Host... 83
        Geographical distribution 83
        Literature 84
        Historical review 84
        Conclusions.. 84
  (16) *Tænia crucigera* Nitzsch and Giebel, 1866........................ 85
        Host... 85
        Geographical distribution 85
        Literature 85
  (17) *Tænia capreoli* Viborg, 1795.................................... 86
  (18) *Tænia capræ* Rud., 1810... 86
Part V.   Life history 87
Part VI.   Conclusions.. 88
Part VII. Compendium of species arranged according to their hosts......... 90–91
Part VIII. Bibliography of adult Cestodes of cattle and sheep............... 92–96
        Addenda 97–101
Index to specific names.. 103

LIST OF ILLUSTRATIONS.

		Page.
Plate I.	*Moniezia planissima*	104
II.	*Moniezia planissima–Moniezia Benedeni*	106
III.	*Moniezia planissima*	108
IV.	*Moniezia Neumanni*	110
V.	*Moniezia expansa–Moniezia denticulata* (Rudolphi's original types)	112
VI.	*Moniezia expansa*	114
VII.	*Moniezia oblongiceps–Thysanosoma Giardi–Tænia marmotæ*	116
VIII.	*Moniezia trigonophora*	118
IX.	*Moniezia trigonophora*	120
X.	*Moniezia alba*	122
XI.	*Thysanosoma actinioides*	124
XII.	*Thysanosoma Giardi*	126
XIII.	*Thysanosoma Giardi*	128
XIV.	*Stilesia globipunctata*	130
XV.	*Stilesia centripunctata*	132
XVI.	*Moniezia, species inquirendæ*	134

LETTER OF TRANSMITTAL.

U. S. Department of Agriculture,
Bureau of Animal Industry,
Washington, D. C., September 15, 1893.

Sir: I have the honor to transmit herewith a report covering "A Revision of the Adult Cestodes of Cattle, Sheep, and Allied Animals." It has long been known that herbivorous animals were subject to infection by tapeworms, but it was not known exactly how many different species were to be found, nor has it heretofore been definitely known what particular species were present in America. Since it may be assumed that every separate species of tapeworm found in animals has a separate source of infection, it has been deemed important to obtain definite data for the determination of the various species already described and to give descriptions of the new forms found in this country. With this end in view, the original type specimens of European species were re-studied according to the more modern scientific methods before the American forms were investigated. The manuscript herewith transmitted covers the results of a more thorough and extensive study of the tapeworms of cattle and sheep than has ever before been attempted, and places the forms mentioned on a scientific foundation.

Very respectfully,

D. E. Salmon,
Chief of Bureau.

Hon. J. Sterling Morton,
Secretary of Agriculture.

9

LETTER OF SUBMITTAL.

U. S. DEPARTMENT OF AGRICULTURE,
BUREAU OF ANIMAL INDUSTRY,
Washington, D. C., April 18, 1893.

SIR: In accordance with instructions received from you to prepare a report on the animal parasites of cattle, I began the work in September, 1891. In the spring of 1892 I took up the tapeworms of cattle, and found that the various diagnoses of the species given by different authors were so contradictory that it was almost impossible to determine a single species with certainty. I was convinced that it was useless to write anything on the subject for my report until the entire group of tapeworms found in cattle and allied animals was subjected to a thorough anatomical study in order to obtain definite characters for the differentiation of the various species.

During the course of the anatomical study then begun I obtained a number of interesting results which are too technical to publish in the biennial report of this Bureau, and yet too valuable to zoölogists and veterinarians and too important theoretically to leave unpublished. These results, therefore, have been included in this manuscript, which I offer for publication as a special bulletin.

In this study I have confined myself almost entirely to the microscopic anatomy of those organs which would be of systematic value. No pretensions are made in this publication to treating the worms histologically, although here and there I have mentioned a few histological obervations.

As it was impossible to determine the various species with certainty from their present diagnoses, I obtained the original type specimens of a number of the parasites in question, and have made my specific determinations by comparing the worms in the Bureau collection directly with the type specimens. I have further received from various European specialists a number of tapeworms which I have compared with our American forms, and it is a pleasant duty to acknowledge the kindness of these foreign specialists in sending me the specimens referred to. To Geheimrath Prof. Dr. Karl Möbius, Director of the Berlin Museum, I am indebted for segments of the original *Tænia expansa* and *Tænia denticulata* described by Rudolphi in 1810; to Prof. L. G. Neumann, of the National Veterinary School at Toulouse, France, I am indebted for specimens of *Moniezia planissima, M. alba, M. trigonophora,*

11

T. Giardi, T. centripunctata, and *T. globipunctata;* to Prof. Moniez, of the Medical Faculty of Lille, France, I am indebted for the original material of the parasites described by him under the names: *T. Giardi, T. Benedeni, T. Neumanni*, and *T. nullicollis*, as well as for specimens of other French tapeworms; to Prof. Railliet, of the Veterinary School of Alfort, France, I am indebted for specimens of *M. planissima;* to Prof. R. Blanchard, of Paris, France, I am indebted for specimens of *Tænia marmotæ*, which I wished to compare with the above-mentioned forms found in cattle and sheep; to Dr. von Marenzeller, of Vienna, I am indebted for a specimen of *M. Benedeni;* to Prof. Perroncito, of Turin, Italy, I am indebted for three strobilæ of his original material from which he described *Tænia alba;* to Dr. G. M. Giles, of Sanawar, Punjab, India, I am indebted for specimens of *Tænia globipunctata.*

Besides the above-mentioned specimens I have had at my disposal the entire Bureau collection of tapeworms of cattle and sheep, over 800 in number, collected by Dr. Curtice, Dr. Hassall, and myself; also a number of specimens from Dr. Hassall's private collection, made in England, and from my own collection of parasites, made in France, Germany, Austria, and America.

I wish to acknowledge, also, the services of my assistant, Dr. Hassall, who has made about 2,000 microscopic slides for my use.

For that portion of this bulletin which deals with new species and for the bibliography at the end of this paper, both Dr. Hassall and myself are to be considered responsible, while I alone am responsible for the statements made in regard to the species already known.

According to the rules passed by the International Zoölogical Congress, Paris, 1889, the law of priority must be strictly adhered to in determining the generic and specific names of animals. Most zoölogists have adopted this rule, although it occasionally causes inconvenience for the time being. In a former paper I objected to applying the rule in all cases, stating, as a particular case, that we should speak of *Pentastomum proboscideum* rather than revert to the specific name *crotali.* At present, however, I wish to withdraw from that position, for I am thoroughly convinced that we can never obtain a fixed nomenclature until the "law of priority" and the rule "Once a synonym always a synonym" are strictly enforced according to the rules laid down by the International Congress. I am hence in favor of speaking of *Linguatula* rather than *Pentastomum*, and if the genus *Linguatula* is divided we are obliged to accept the genera, *Linguatula* and *Porocephalus—Porocephalus crotali*, instead of *Pentastomum proboscideum.**

---

*We should in that case speak of *Porocephalus constrictus, Por. annulosus, Por. polyzonus, Por. subuliferus, Por. moniliformis, Por. megastomus, Por. oxycephalus*, etc. It is not at all improbable that *Porocephalus armillatus* (*Linguatula armillata* Wyman, 1845) will prove to be identical with *Porocephalus polyzonus*, in which case Wyman's specific name must be accepted. I hope later to compare the type specimens of some of these forms in order to straighten out the synonymy, which seems at present to be very uncertain.

In the present paper I have held strictly to these rules, and used the terms *actinioides* and *Giardi* as specific names, instead of *fimbriata* and *ovilla* in speaking of two of the worms found in sheep.

I am well aware of the fact that in following out these rules I am doing that which, although indorsed by most zoölogists, is disapproved of by many helminthologists, who contend that we should accept the specific name which we find in general use. These authors, however, lose sight of the fact that many animals have one name in general use in one country, while another name is generally used in another country. Thus *Pentastomum* is in general use in Germany, while *Linguatula* is in more general use in France, England, and America. Authors are generally best acquainted with the text-books of their own country, and will naturally assume that the name "in general use" in their own country is the name to be followed. Thus Germans would hold to *Pentastomum*, while French, English, and American authors would hold to *Linguatula*.

The only way to establish an international nomenclature in medical zoölogy, as well as in other branches of biology, is to enforce the law of priority.

In the adult tapeworms of cattle and allied animals I recognize several distinct groups, and in the description given below the parasites are classed together in four groups:

(1) Blanchard's genus *Moniezia*, in which the species naturally fall into three subdivisions:

(a) The *Planissima group*, characterized by the linear arrangement of the interproglottidal glands (*M. planissima, M. Benedeni, M. Neumanni*).

(b) The *Expansa group*, characterized by the saccular arrangement of the interproglottidal glands (*M. expansa, M. oblongiceps,* and *M. trigonophora*).

(c) The *Denticulata group*, which contains forms in which the interproglottidal glands are absent (*M. denticulata* and *M. alba*).

(2) The genus *Thysanosoma* Diesing. This genus is characterized by the presence of a single uterus with ascon-spore shaped egg-sacs. Further, while in the genus *Moniezia* the genital canals pass dorsally of the longitudinal canals, in the genus *Thysanosoma* they pass between the longitudinal canals.

(3) Railliet's new genus *Stilesia*, based upon *Tænia globipunctata*. Provisionally the species *T. centripunctata* may be placed in this genus, but I believe that further study may result in the formation of a separate genus for that worm.

(4) Owing to our lack of knowledge of the anatomical details of *M. nullicollis, T. Vogti, T. crucigera, T. capreoli,* and *T. capræ*, it is impossible as yet to give these species any definite place in the classification.

Besides the above-mentioned species, the Bureau possesses three forms of *Moniezia*, taken from sheep, at least two of which are probably new species.

Unless otherwise expressed the measurements given are in millimeters (^{mm}).

The cestodes collected by this Bureau for the past eighteen months have been fixed in the following solution: Fifty parts of an aqueous solution of corrosive sublimate + 50 parts of alcohol, 70 per cent + a few drops of glacial acetic acid. The worms were placed in this liquid which had been heated to 45–53° C. The liquid was then allowed to cool for 20 to 60 minutes. The parasites were next washed in running water from 1 to 24 hours, and passed through 30 per cent, 50 per cent, 70 per cent, 95 per cent, and absolute alcohol.

The preparations have been colored in alcoholic hydrochloric acid-carmine after Grenacher or hæmatoxylin, and finally mounted in Canada balsam. Most of the specimens collected by the Bureau prior to July, 1891, were simply preserved in alcohol, and, with few exceptions, have been of very little use in preparing this bulletin. It is also evident that the specimens we have received from European zoölogists have been killed in alcohol.

Anterior in this paper signifies cephalad.

Posterior signifies caudad.

Proximal refers to the direction toward the head or median line.

Distal refers to the direction from the head or median line.

Ventral refers to the flat surface of the worms, containing the large longitudinal canals and female organs.

Dorsal refers to the opposite surface.

Lateral refers to the direction toward the margins upon which the genital pores are found.

Respectfully submitted.

CHARLES W. STILES,
Zoölogist, Bureau of Animal Industry.

Dr. D. E. SALMON,
Chief of the Bureau of Animal Industry.

A REVISION OF THE ADULT CESTODES OF CATTLE, SHEEP, AND ALLIED ANIMALS.

By C. W. STILES, PH. D., and ALBERT HASSALL, M. R. C. V. S.

PART I.

MONIEZIA R. Blanchard, 1891.

A. Planissima Group. (*M. planissima, M. Benedeni*, and *M. Neumanni.*)

Interproglottidal glands linear and not grouped around sacs.

(1) Moniezia planissima Stiles and Hassall, 1892.

[Plates I, II, Figs 1-6; Plate III.]

Synonymy.—*Tænia expansa, ex parte* of various authors.
Hosts.—Sheep (*Ovis aries*); cattle (*Bos taurus*); small intestine.
Geographical distribution.—United States of America: Chicago, Ill., and Washington,
    D. C. (Hassall, Stiles, and Curtice); Fairfax County, Virginia (Norgaard).
    France: Paris (Railliet, of Alfort, Stiles); Toulouse (Neumann).

LITERATURE.

(Probably all references to *Tænia expansa* 20-25^{mm} broad.)

(1) C. W. STILES. Notes sur les Parasites—14: Sur le *Tænia expansa* Rudolphi;
    Compt. rend. de la Soc. d. Biol. Paris, 1892. Pp. 665–666.
(2) ———— Bemerkungen über Parasiten—17: Ueber die topographische Anato-
    mie des Gefasssystems in der Familie *Tæniadæ;* Centralblatt für Bakteriol-
    ogie und Parasitenkunde. 1893. Bd. XIII, p. 457–465, Fig. 3.

HISTORICAL REVIEW.

Upon studying the tapeworms of cattle and sheep we became convinced that *Moniezia expansa* contained more than one species, and diagnosing the proper *M. expansa* by a few segments and the scolex of Rudolphi's original material, we separated from it *Moniezia planissima.* Stiles found this species at Paris, and Hassall and Stiles found it at Washington, D. C. We have also found several segments of this worm in the material collected by Dr. Curtice, under the label "*Tænia expansa,*" and Dr. Norgaard has recently found it in Virginia. Two of the best known medical zoölogists of France (Railliet and Neumann) have also sent us specimens of this species under the label "*Tænia expansa.*"

The only publications on this worm, as a distinct species, are the short preliminary notes by Stiles (1, 2). The present account of the anatomy is the paper anticipated by those preliminary papers.

15

ANATOMY.

General appearance.—When fresh the strobila is yellowish, but when preserved it becomes whitish. The worm is remarkably flat and broad much more so than *M. expansa.* In the best specimens at our disposal the segments are slightly contracted. In a few specimens in the Bureau collection they are not contracted, the segments being slightly longer and narrower. In the following description, however, we have used the worms which are slightly contracted. Nevertheless, we have given a figure of some uncontracted segments, so that the difference may be readily seen and no confusion arise.

Fig. 1, Plate I, represents a strobila, natural size, slightly contracted. The segments increase in breadth very rapidly. It will be noticed that the segments are very regular, although wedged segments, such as have been described in other worms, are occasionally found. The posterior flap of the segment, overlapping the next following segment, is very distinct. The cuticle is very smooth.

Scolex.—The heads of the various preserved specimens in our possession vary greatly in shape and size. They have the appearance presented in Figs. 2, 2a and 2b, Plate I. When viewed from the apex the head is nearly square. The suckers are at the four corners and turned slightly towards the front. In press preparations, viewed from the ventral or dorsal surface, we notice a plane running from the anterior portion backwards and upwards, i. e., the posterior edge of this plane is the point of the highest focus, and situated 0.56 ^{mm} from the anterior extremity. A head examined in alcohol measured 0.8^{mm} broad. The suckers were strong, 0.296 ^{mm} in diameter, the muscular wall 0.064^{mm} thick; cavity of the sucker round; the opening somewhat oblong and directed diagonally toward the front. The constriction (0.7^{mm} from the anterior extremity) is 0.664^{mm} broad. Then follows an unsegmented portion (neck) 0.8^{mm} long, at the end of which the first signs of strobilization are visible. In some specimens the neck appears short and thick, in others very long and thin (contraction). In the neck two longitudinal canals are visible on each side.

In the examination of preserved material too much confidence must not be laid upon the size of the head, as it may contract or shrink in preparation. Press preparations also give more or less artificial pictures. Thus the following measurements were found in preserved material and balsam preparations:

Diameter of head.	Posterior edge of plane from anterior point.	Diameter of suckers.	Muscular wall of suckers.	Distance of constriction from anterior extremity.	Diameter of constriction back of the head.	Strobilization begins back of constriction.
mm.	mm.	mm.	mm.	mm.	mm.	mm.
1.. 0.40	0.528	0.30	0.072	0.8	0.312	0.8
2.. 0.548	0.326	0.288	......	0.480	0.352	0.64
3.. 0.448	......	0.192	0.072	0.4	0.216	0.8
4.. 0.67	......	0.256	0.064	0.56	0.496	1.48

Most of these variations are due to contraction. The head of fresh specimens generally measures 0.7–0.9^{mm}. The segments are at first extremely short and indistinct, but they rapidly become much broader, longer, and sharper in their outline. Seven to 12^{mm} from the head (segment *ca.* 150) the segments measure 0.062–0.112^{mm} long by 1.5–2^{mm} broad; they are very thin. At this point (Plate I, Fig. 3) we find the first trace of genital organs in the form of round clumps of tissue, 48 μ in diameter, in the lateral portion of the median field (just inside the longitudinal canals); two anlagen* are present in each segment.

Segments 70^{mm} from the head measure 4^{mm} broad by 0.3–0.55^{mm} long (Plate I, Fig. 4). The genital pores have not yet pierced the edge of the segment. The anlagen of the genital canals are still solid and extend from the lateral margin of the anterior half, a distance of about 0.3^{mm} towards the median field, crossing the longitudinal canals and ending in a double knob. The anterior portion of the knob belongs to the male genitalia, the posterior to the female genitalia. No testicles are visible.

The distal border of every segment is very slightly wavy and overlaps the proximal border of the next following segment about 0.032^{mm}. The juncture of the segments is differentiated in the median field to the extent of showing a linear group of cells each side of the median line; this organ will be described more fully below.

The longitudinal canals, which were of nearly the same diameter in the head, have changed in position and size. The ventral canal has a lumen of about 0.1^{mm}, and is nearer the margin of the segment; the dorsal canal is but 0.05^{mm} broad, and lies on the median side of the ventral canal, at the same time slightly dorsal of the latter.

The testicles first appear in segments about 100^{mm} from the head. At first they are arranged in two triangles pointing towards the median line, as in *M. trigonophora* (see p. 38), but they very soon lose this arrangement (see below).

Segments 14^{cm} from the head measure 8^{mm} broad and 0.4^{mm} long. The genital pore is distinct and the male and female canals are very plain. The genital pore is in the anterior half of the edge of the segment. The vas deferens is very slightly convoluted; the testicles have increased in number and they are nearly as numerous in the median as in the submedian fields. None were found in the lateral fields.

Of the female organs one can distinguish vagina, receptaculum seminis, shell-gland, vitellogene gland, ovary, and several small canals which will be considered below. The uteri are not visible. In the

*The word "anlage"—plural "anlagen"—is now being adopted by a number of American zoölogists to signify the *incipient rudiment* of an organ. This is the meaning given to the word in this paper. Prof. E. L. Mark has recently introduced the term "fundament" as equivalent to the German "Aulage."

median field each segment overlaps the one next following by 0.176ᵐᵐ; and at the juncture of every two segments there is a differentiation in the median field (about 0.68ᵐᵐ each side of the median line) which colors very dark in acid carmine. Sacs such as we find in *M. expansa* (see p. 32) are, however, nowhere developed. The arrangement of the various organs can be especially well seen in segments 180-400ᵐᵐ from the head, measuring 9-13.5ᵐᵐ in breadth by 0.67-0.96ᵐᵐ long by 0.17-0.26ᵐᵐ thick.

The segments are quadrate in form (Plate II, Fig. 4). The posterior flap of each segment overlaps the anterior portion of the next following segment by 0.192ᵐᵐ (Plate I, Fig. 5). The first thing which strikes the eye on the stained specimen is the deeply colored line at the juncture of the segments, running parallel to the posterior edge and measuring about 1.9ᵐᵐ long by 48 μ wide. Upon examining this in sections with a high power it is found to be distinctly cellular in structure. On frontal sections (Plate II, Fig. 2) the cells are arranged irregularly in groups. This grouping is more or less artificial, and caused by the longitudinal muscles. The cells are drawn out at one end, which inclines toward the nearest muscle-fiber. The round nuclei (4 μ) are very distinct and the nucleolus (2 μ) is very sharply defined. Generally only one dark spot is found in the nucleus, but occasionally two are found. In Plate III, Fig. 2, it will be seen that these cells surround the border between two succeeding segments, and that they are directed towards the cuticle. They have the same appearance as the cells surrounding the interproglottidal sacs in *M. expansa*, and probably represent a low stage in the phylogenetic development of those organs. The grouping which, as stated, is in this case irregular and apparently dependent on the position of the longitudinal muscle fibers, reaches its highest development around the sacs in the species of the *Expansa group*.

The genital pores are in the lateral margin, invariably in the anterior half of the segment (Plate II, Fig. 4, and Plate III, Fig. 3). The male and female openings present the following constant topographical relations: On the right hand side of the segment the vagina is ventral, while the cirrus lies dorsal in the same transverse plane; on the left side the position is reversed so that the vagina is dorsal and the cirrus is ventral in the same transverse plane (Plate II, Fig. 3, 4). This arrangement was found constant in every segment examined.

The vagina leads toward the female glands, which lie just median of the longitudinal canals, while the vas deferens extends from the cirrus across (dorsally) the female glands to the testicles, which are scattered throughout the median field (Plate II, Fig 4).

Male genitalia.—The cirrus-pouch extends straight towards the median line of the body. It is 0.28ᵐᵐ long, 0.096ᵐᵐ in diameter (Plate II, Figs. 3, 4, and Plate III, Fig 4). The inverted cirrus runs through its center, forming but comparatively few convolutions. The layers of the cirrus-pouch, beginning at the center, are as follows: (1)

lumen; (2) ciliary projections extending into the lumen. Zschokke states that this is of cellular nature in *Tænia expansa*, but we were unable to convince ourselves of the correctness of this statement as applied to *M. planissima*. We saw in sections a number of nuclei which were apparently in this layer (Plate III, Fig. 4) but, which upon further study proved to be in the next following layer; (3) a more or less homogenous layer containing numerous nuclei, hence of cellular origin. In (Plate III, Fig 4) some of these nuclei appear as if they were in layer 2, but the circular muscles prove that the section was not exactly straight; 2 and 3 are probably to be looked upon as a single layer, or more strictly speaking, as arising from a single layer of cells; (4) circular muscles; (5) longitudinal muscles, *i. e.*, longitudinal in reference to the shape of the cirrus-pouch, not in reference to the worm; (6) cellular parenchyma with large nuclei; (7) circular muscles with nuclei which stain very dark. This layer is much thicker in the proximal than in the distal portion of the pouch; (8) longitudinal muscles.

For the study of layers 7 and 8 this species is not very good. In cross sections of the pouch the circular muscles (Plate III, Fig. 6) are very evident, but it is only with the greatest difficulty that the longitudinal fibers are seen, while in longitudinal sections the reverse is the case, except that the circular fibers of the proximal portion come plainly into view.

Upon emerging from the median end of the cirrus-pouch the male canal immediately increases in diameter, forms a number of convolutions, and crosses into the median field on the *dorsal* side of the longitudinal canals; most of the convolutions lie anterior to the vagina. Arriving at the ovary it decreases in diameter and crosses the latter dorsally.

Testicles.—At this stage there are 400 to 600 testicles scattered through the entire median field, forming a more or less regular quadrangle. There is generally no portion in the median line where the testicles are absent, such as is seen in *M. trigonophora*. On cross section they are seen to lie in the center portion of the segment, *i. e.*, between the transverse muscles.

Female genitalia.—In a depression directly dorsal (left side, ventral right side) of the cirrus lies the vulva (Plate II, Fig. 3), from which the vagina runs almost straight toward the median field of the segment. The layers of the vagina (Plate III, Figs. 4, 5) correspond very nearly to those of the cirrus, *i. e.* (1) lumen; (2) ciliary projections; (3) homogenous layer with nuclei; (4) longitudinal fibers (almost imperceptible); (5) circular fibers, which are extremely thin except at the beginning of the vagina, where they form a strong sphincter capable of completely closing the entrance; and (6) a distinct cellular layer similar to that recognized by Zschokke in *T. expansa*.

It will be noticed that in the cirrus the circular muscles (4) are surrounded by the longitudinal fibers (5), while in the vagina the reverse is the case. This has already been noticed by other authors, Zschokke for instance, in other species (*T. expansa*).

The vagina runs toward the median field and passes the longitudinal canals on their dorsal side, retaining a nearly uniform diameter for about 0.72^{mm}. Then it suddenly enlarges (Plate III, Fig. 3) into a short bulb, which is followed by a constricted portion 24 μ long. The latter then empties into the receptaculum seminis, which measures 0.44^{mm} long by 0.096^{mm} in diameter and lies on the dorsal side of the ovary. The histology of the receptaculum seminis is entirely different from that of the vagina, there being an inner epithelium with an outer basement membrane. From this point the topography of the canals can be made out only by a patient study of numerous sections. Our studies on sections in the three directions have led us to the following results (Plate II, Fig. 1 and Plate III, Figs. 1–3):

From the ventral side of the receptaculum seminis, a short distance from its median extremity, a canal takes its origin; this canal runs about 0.132^{mm} in a curved direction, the convexity being ventrad and slightly distal; it is made up of a high interior epithelium (nuclei 3 μ) with a basement membrane. This canal (oviduct) in which, as we shall see later, the ova come in contact with the spermatozoa (hence " fertilization canal"), leads from the ovary, from which it is separated by a short constriction provided with circular fibers. Thirty-two μ from the origin of the above-described canal another canal branches off from it dorsally, makes one or more convolutions, and enters the shell gland; in about the center of the latter it divides into two canals, one of which passes straight through the gland in a ventro-distal direction and dilates suddenly into the calyx of the vitellogene gland. This latter canal is the vitello-duct. The other canal runs from the center of the gland dorso-anteriorly, makes several convolutions both in the shell-gland and after it leaves it, then runs to the uterus. The relations of these canals are given in Plate II, Fig. 1, and Plate III, Fig. 1, which are slightly diagrammatic.

Ovary.—The ovary is nearly renal in form, varying slightly according to the contraction of the segment (Plate III, Fig. 3). When the segments are expanded, the ovary, shell-gland, and vitellogene gland together form a circular body, but when the segment is more or less contracted, this circular body becomes lengthened in its lateral diameter and shortened in its longitudinal diameter. The ovary in Plate III, Fig. 3, measures 1.12^{mm} in its lateral diameter. Coloring liquids differentiate the organ into two distinct parts; a peripheral portion (cortex) which stains very dark and represents the ovarial tubes, and a central portion (calyx) which stains more lightly. All the tubes lead into this central portion from which the oviduct runs.

The vitellogene gland (Plate II, Fig. 1, and Plate III, Figs. 1–3) lies on the posterior (distal) side of the ovary, and is nearly inclosed by the projecting sides of the latter. It measures 0.4^{mm} by 0.24^{mm}. In preparations this organ appears much darker than the ovary, owing to the fact that the nuclei are of a darker color, slightly smaller, and more crowded. The gland is composed of tubes which lead into a common calyx from which the vitello-duct takes its origin. While the ovary consists of long, thin tubes, the vitellogene gland is more acinus-like in its character.

Shell-gland (Plate II, Fig. 1, and Plate III, Figs. 1–3).—Between the ovary and vitellogene gland, and on the dorsal side of the latter, is situated an irregularly shaped organ which is known as the shell-gland. The point where the vitello-duct and oviduct come together (oötyp) forms approximately the center of the gland. The nuclei of the gland (5 μ) are very similar to those of the ovary, and take on about the same stain.

Uterus.—No connection can be traced between the uteri-anlagen of the two sides of the segment, hence we have originally (ontogenetically) two uteri, corresponding to the two sets of organs.

Excretory canals.—The ventral longitudinal canal lies laterally of the dorsal canal and slightly ventrally. Its lumen is flattened dorso-ventrally (0.2^{mm} by 0.016^{mm}) and is extremely irregular; its cuticular lining is very thin. In the posterior portion of each segment it is connected with the corresponding canal of the other side by the transverse canal. The dorsal canal is smaller (0.064^{mm} by 0.008^{mm}); the cuticular lining is thicker than that of the ventral canal. Cells are grouped more or less regularly around the dorsal canal, but we can hardly look upon them as forming a distinct epithelium such as Zschokke described in his article on *T. expansa.*

The longitudinal nerves can be distinguished on the lateral side of the lacunes. They cross the genital canals ventrally. From this point to the end of the strobila the segments continue to increase considerably in breadth but very slightly in length. Segments 60^{cm} from the head measured 14^{mm} by 1^{mm}; 80^{cm} from the head, 16^{mm} by 1^{mm}; 1^{m} from the head, 18^{mm} by 1.5^{mm}, and 1.25^{m} from the head, 20^{mm} by 1.75^{mm}. In one worm segments 26^{mm} broad were found. The last few segments show a disposition to become slightly longer and narrower.

As the uteri increase in size they make numerous folds which completely fill up the median field and extend into the lateral fields, crossing the longitudinal canals dorsally. In the median line the two uteri come together in such a way that it is impossible to distinguish one from the other (Plate II, Fig. 5).

In the meantime the other genital organs atrophy, the testicles being the first to disappear; then the ovary, vitellogene gland, and shell-gland fade; the receptaculum seminis, vagina, and cirrus-pouch persist the longest (Plate II, Fig. 5).

Eggs.—The ova are more or less globular, measuring 63 μ in diameter; by reciprocal pressure they become cuboid. The embryo, which measures 16–18 μ in diameter and is provided with the six characteristic hooks 9 μ long, is surrounded by a pyriform apparatus. The bulb of the apparatus—*i. e.*, that part directly surrounding the embryo—measures 20 μ in diameter. The horns are 24 μ long, appear to be hollow, and end in a disk 15 μ in diameter. When viewed *en face* the disk is nearly round. In the center are seen the ends of the two horns, each 3 μ in diameter, and several—generally four—round nuclear bodies (3 μ), each containing one or two very refringent dots. These four bodies have every appearance of being true nuclei. When treated with caustic soda the disk frequently assumed the fringed appearance which Curtice figures in *Tœnia expansa*.

There is a thin membrane inside the bulb of the pyriform apparatus, and at a point near the horns this membrane was invaginated toward the embryo (Plate II, Fig. 6). This observation was made upon preserved material, so that it is not impossible that this is to be explained by a contraction of the embryo. The horns are surrounded by a granular fatty substance.

SPECIFIC DIAGNOSIS.

From the above anatomical description we feel justified in giving the following specific diagnosis:

Moniezia planissima S. & H., 1892.—Adult strobila 1 to 2ᵐ long; yellowish. Head quadrate, 0.4–0.9ᵐᵐ broad; suckers at the four corners and directed forwards (diameter 0.25ᵐᵐ); dorsal and ventral surfaces usually show a distinct plane running across the suckers and ending about 0.5ᵐᵐ from the anterior extremity (balsam preparations). Neck broad, 0.6–1.5ᵐᵐ long or very narrow and long. First traces of genital organs 7–12ᵐᵐ from the head (segment *ca.* 150). The segments are at first extremely short; always much broader than long and longer than thick; ripe segments 12–26ᵐᵐ broad by 1–1.75ᵐᵐ long, generally thin. Interproglottidal glands linear, large and very distinct. Genital pores double, situated in the anterior portion of the lateral margin; vagina and cirrus on the same transverse plane—on the right, vagina ventral, cirrus dorsal; left side vagina dorsal, cirrus ventral; genital canals pass dorsally of the longitudinal canals and nerves. Testicles begin about 10ᶜᵐ from the head, at first arranged in two triangles; 400–600 testicles present in ripe segments and generally as numerous in the median line as elsewhere, so that the testicles of each side cannot always be distinctly separated. Uterine folds enter the lateral fields. Eggs 63 μ in diameter; bulb of pyriform apparatus 20 μ; horns 24 μ; embryo 16–18 μ; hooks 9 μ. (See also generic diagnosis, p. 54).

Type specimen is in the Bureau of Animal Industry; typical specimens will be sent to museums and specialists exchanging with this Bureau.   S. & H.

(2) Moniezia Benedeni (Moniez, 1879) R. Bl., 1891.

[Plate II, Figs. 7, 8.]

Synonymy.—*Tœnia Benedeni* Moniez, 1879; *Moniezia Benedeni* (M.) R. Bl., 1891; *Tœnia denticulata* Neumann, 1892 (fig. of scolex.)

Hosts.—Domestic sheep (*Ovis aries*), Moniez; cattle (*Bos taurus*), Moniez, v. Marenzeller.

Geographical distribution.—France (at Lille, Moniez); Austria (at Vienna, v. Marenzeller).

23

LITERATURE.

(1) MONIEZ, R. Note sur deux espèces nouvelles de Tænias inermes (*Tænia Vogti* et *T. Benedeni*); Bulletin scientif. du Department du Nord, 1879, 2 sér., T. 2, p. 163.

(2) RAILLIET, A. Éléments de Zoologie Médicale et Agricole, 1886, p. 261, *Tænia Benedeni*.

(3) NEUMANN, L. G. Traité des Maladies parasitaires non-microbiennes des animaux domestiques, 1st ed., 1888, p. 383, *Tænia Benedeni*.

(4) BLANCHARD, R. Sur les Helminthes des Primates anthropoïdes; Mém. d. l. Soc. Zool. d. France, 1891, p. 187, footnote, *Moniezia Benedeni*.

(5) MONIEZ, R. Notes sur les Helminthes. VI, 1; Revue Biologique du Nord d. l. France, 1891 (Extr. pp. 14–16), *Moniezia Benedeni*.

(6) BLANCHARD, R. Notices helminthologiques. 2. sér., 7; Mém. d. l. Soc. Zool. d. France, 1891, p. 444, *M. Benedeni*.

(7) NEUMANN. Traité, etc., 2 ed., 1892, p. 408, *T. Benedeni.*—Engl. Transl., p. 418, *T. Benedeni*.

HISTORICAL REVIEW.

This species was first described by Moniez (1) as a worm of 4ᵐ or more long; head large in proportion to the neck; last segments more than a centimeter broad; longitudinal muscles of one surface developed more than those of the other surface. The parasites were found in the intestine of sheep at Lille, France. The above diagnosis is also given by Railliet (2). Neumann (3) adds that the head is small, globular, and followed by a filiform portion 4–5ᶜᵐ long, in which the segments are visible; the eggs are 75–80 μ, and show a pyriform apparatus. R. Blanchard (4) then states in a footnote that he includes this species in his new genus *Moniezia*. This publication was immediately followed by one from Moniez (5), in which he accepts Blanchard's classification and adds the following in regard to *M. Benedeni*: "The worm is remarkable on account of the thickness and breadth of the segments; in contraction the latter attain 12ᵐᵐ in breadth, 3ᵐᵐ in length, and 2ᵐᵐ or more in thickness; the head is more than 1ᵐᵐ thick; the suckers are strong, prominent, 0.45ᵐᵐ in diameter, and touch each other; same form found in seven individuals; neck 2.5ᵐᵐ long; form of segments the same through entire length; found in cattle as well as sheep. Ova 80–85 μ, pyriform apparatus 18 μ, embryo 15 μ." In Blanchard's second publication (6) the worm is simply mentioned as belonging to the genus *Moniezia*. In the second edition of Neumann's work (7) the specific diagnosis is not enlarged, but a figure is given of a head of the worm which the author considers to be *T. Benedeni*. In response to a request both Moniez and Neumann kindly forwarded specimens which they included in this species. Neumann's specimen, however, agrees perfectly with *Moniezia trigonophora*.

Moniez's specimen bears the label "Mouton 31 (? indistinct—c. w. s.), 79, i, prep. xx. le 2 gros. fragm.=? du même." As he described this parasite in 1879, and as he was asked for his original type, it is assumed that the "79" refers to the year, and that the sample sent is some of his original material; this specimen forms the basis of the present description. Moniez's interrogation point on the label signified that he suspected there was more than one species contained in the bottle. His suspicions proved to be perfectly correct, for besides the *M. Benedeni*, which was easily recognized from his description of the head, segments of *M. expansa* and *Th. Giardi* were found. A tapeworm labeled "*T. expansa*" was received from Dr. von Marenzeller, of Vienna, which agrees with Moniez's type of *M. Benedeni*.

ANATOMY.

External appearance.—Plate II, Fig. 7, represents a strobila (natural size) which was received from Moniez. The animal is considerably contracted. It will be noticed that the head is quite large, the neck is quite stout, the segments are distinct, thick and much broader than long. Fig. 7a gives two enlarged segments, which show a peculiar checkered appearance of the cuticle; this is common to all the large segments. The strobila is very opaque.

Scolex.—Figs. 8, 8a agree perfectly with Moniez's description (5) of the head. Upon comparing these figures with Neumann's figure of the scolex of *T. denticulata* (7, Fig. 183), it will be noticed that they are almost identical. Prof. Neumann was kind enough to send 4 strobilæ of the worm which he looked upon as *T. Benedeni*, and they do not agree with Moniez's type, but, as stated above, with *Moniezia trigonophora*, so that we are forced to assume that Neumann's diagnosis of *M. Benedeni* must be dropped, and that his *T. denticulata* (partim) is probably identical with Moniez's *M. Benedeni.*

The head of Moniez's *M. Benedeni*, in the possession of the writer, is square, 0.96^{mm} by 0.96^{mm}; length 0.48^{mm}. Two furrows cross the anterior surface of the head and extend around on the four sides; these furrows are connected by furrows posterior to the suckers, thus separating the latter from the strobila. This arrangement of furrows gives the head a lobed appearance, each of the 4 lobes representing a large sucker 0.48^{mm} in diameter; the opening of the suckers must of course vary with the state of contraction. In Figs. 8, 8a the opening is small and circular. The specimen of this species is so contracted that it would be altogether too hazardous to make many statements in regard to its internal anatomy. It can, however, be stated that the interproglottidal glands are *extremely indistinct* (absent?), without sacs, but *appear* to be linear, as in *M. planissima.* There appears to be no reason, however, for uniting these two species, since the heads and strobilæ are so totally different. Furthermore, Moniez states that the contracted segments measure 12^{mm} broad by 3^{mm} long by 2^{mm} thick, which measurements do not agree with those of *M. planissima.* The general topographical relations of the vagina, cirrus, genital canals, genital glands, longitudinal canals and nerves are the same in this species as in *M. planissima;* the arrangement of the testes, however, was not observed.

Ova.—As stated above, Moniez gives the measurement of the ova as 80–85 μ, of the embryo 15 μ.

SPECIFIC DIAGNOSIS.

From our present knowledge of this parasite we may accept the following as a provisional diagnosis:

Moniezia Benedeni (Moniez, 1879) R. Bl., 1891.—Strobila attains a length of 4^m; head 1^{mm} or more large; obtuse, 4-lobed, suckers directed diagonally forwards, opening circular. Neck present, slightly narrower than the head, about 2-2.5^{mm} long. Segments always broader than long; contracted ripe segments attain the measurement of 12^{mm} broad by 3^{mm} long by 2^{mm} thick. Topography of longitudinal canals, genital canals, and female glands the same as in *M. planissima;* testicles? Interproglottidal glands *extremely indistinct* [absent?], without sacs. Ova 80–85 μ in diameter; embryo 15 μ; pyriform apparatus 18 μ.

Type specimens are in the possession of Prof. Moniez, Lille, France, and C. W. Stiles, Washington, D. C. One very typical specimen is in the K. K. Museum in Vienna, Austria.

C. W. S.

(3) Moniezia Neumanni Moniez, 1891.

[Plate IV, Figs. 1-5.]

Host.—Sheep (Moniez).
Geographical distribution.—France (Lille, by Moniez).

LITERATURE.

(1) Moniez, R. Notes sur les Helminthes, vi, 2; Revue Biologique du Nord de la France, 1891, t. iv, 1 page.

HISTORICAL REVIEW.

Moniez gave a very short description of this worm (1), by which it has not been possible to recognize the species. He was, however, so kind as to send specimens. He states (1) that the head has the same dimensions as that of *M. Benedeni,* and is more than 1^{mm} in diameter; suckers agree with those of *M. Benedeni;* while the neck is the same length as in *M. Benedeni* but filiform. The strobila is much shorter (1–2 feet); segments narrower (8^{mm}), thinner, and 1–1.5^{mm} long; the segments which are about to become detached are 6 by 2^{mm}. Ova measure 55–56 μ, embryo 18–21 μ.

ANATOMY.

Only contracted segments (Plate IV, Fig. 1) are at hand, and but little in regard to the anatomy of this species can be given.

Scolex.—The head resembles that of *M. Benedeni* in some respects; it is almost square when viewed *en face* (Plate IV, Fig. 2a), 0.9^{mm} broad; when viewed from the dorsal or ventral surface, the suckers do not appear so distinctly separated from the neck. The openings of the suckers are quite small, situated at the four corners and directed diagonally forward. The segments are all broader than long.

In several segments measuring 1.2^{mm} wide by 0.56^{mm} long (Plate IV, Figs. 3, 4), the genital pore was just on the point of piercing the lat-

eral margin of the segment in its anterior third. The male and female
canals were distinct from each other, but their lumena were very indistinct. The female glands were very small, but could be distinguished.
A large number of testicles were distributed in the median field, and
they were as numerous in the median line as elsewhere. Short linear
interproglottidal glands without sacs were present in the median line.
In several segments (Plate IV, Fig. 5) measuring 4^{mm} wide by 2.2^{mm} long,
the uteri occupied almost the entire segment and were filled with ova.
The other organs were atrophied. The genital pore was in the middle
or anterior half of the segment. In the segments 6.5^{mm} wide no details
were recognized.

SPECIFIC DIAGNOSIS.

The specific diagnosis of this cestode is very unsatisfactory, owing
to our lack of knowledge in regard to important details.

M. Neumanni M., 1891.—Strobila 1½ to 2 feet long, perhaps longer. Head square,
0.9^{mm} or more broad, suckers powerful, same diameter as in *M. Benedeni*, but not
so distinctly lobed as in that species. Neck about 2–2.5^{mm} long, anterior portion
filiform, very thin. Segments narrower and thinner than those of *M. Benedeni;* the
largest segments measure 8^{mm} broad by 1.5^{mm} long; the segments at the posterior end
somewhat narrower and longer, 6^{mm} by 2^{mm}. Topography of longitudinal canals, genital canals, and female glands probably the same as in *M. Benedeni;* testicles arranged
in a quadrangle. Interproglottidal glands small, linear, without sacs. Ova 55–65
μ; embryo 18–21 μ.

Type specimens are in the possession of Prof. Moniez, Lille, France,
Bureau of Animal Industry, and C. W. Stiles, Washington, D. C.

It will be seen from the above that *M. Neumanni* as well as *M. Benedeni* must be subjected to a thorough anatomical study by some one
who can obtain the proper material. The powerful contraction of the
specimens, and the inability to obtain fresh material, is offered as an
apology for not going more into details. See also *M. nullicollis* (p. 83).

C. W. S.

B. Expansa Group. (*M. expansa, M. oblongiceps,* and *M. trigonophora.*)

Characterized by the grouping of the interproglottidal glands around blind sacs.

(4) Moniezia expansa (R., 1810) R. Bl., 1891.

[Plate V, Figs. 1–3; Plate VI.]

Synonymy.—(After Rudolphi) *Tænia vasis nutritiis distinctis* Bloch, 1782; *T. ovina*
Goeze, 1782; *T. ovina* Batsch, 1786; *T. ovina* Schrank, 1788; *T. ovina* Gmelin,
1789; *Halysis ovina* Zeder, 1803—(After Creplin) *T. denticulata* Mayer—(After
Baird, 1853) *Alyselminthus expansus* Blainville—(After Stiles) *T. expansa, ex
parte* of Railliet, Neumann, Perroncito, Curtice, and other recent authors.

Hosts.—Sheep (*Ovis aries*), found by various authors; cattle (*Bos taurus*), found by
various authors; Zebu (*Bos indicus*), found by Perroncito; goats (*Capra
hircus*); Pyrenian tor or ibex (*Capra pyrenaica*); roe deer (*Capreolus caprea*),
reported by Nitzsch; pampas deer (*Cariacus campestris*), see Dies.; brocket
(*Cariacus rufus*), see Dies.; *Cervus Nambi*, see Dies. (collected by Natterer);
Gazella dorcas, reported by Bremser, see Dies.; *Ovibos moschatus*; gemse or
chamois (*Rupicapra tragus*), reported by Bremser.

Geographical distribution.—North America: Chicago, Ill., *Stiles; Washington, D. C., *Curtice, *Hassall, *Stiles; Colorado, Curtice. South America (Brazil, Natterer). Europe: England, Hassall and others; France, *Railliet, *R. Blanchard, Neumann, Moniez, *Stiles, and others; Germany and Austria, Leuckart, Bloch, Goeze, Nitzsch, *Rudolphi, *Zschokke, *Stiles, and many others; Italy, Perroncito and others.

Besides the above, this species is mentioned by numerous other authors in various parts of the world. In all probability some of the cases are those of true *M. expansa*, while in other cases the data given are too incomplete to determine whether the species found was *M. expansa* or some other form. An examination has been made of the specimens of this species collected by those men whose names are starred (*) in the above list. It is very certain that many of the cases of *Taenia expansa* reported in America are either *Th. actinioides*, *M. planissima*, or *M. trigonophora*. In all probability this is the most widely distributed species found in cattle and sheep (See extract from letter by Dr. Giles, p. 74.)

LITERATURE (1810–1892).

(1) Rudolphi. Ent. Hist. nat., ii, 2, 1810, p. 77–79, *Taenia expansa*.
(2) ———— Ent. Syn., 1819, pp. 144, 487.
(3.) Gurlt. Lehrbuch der pathologischen Anatomie der Haussäugethiere, 1831, i, p. 381, x, 1–2.
(4) ————. Hartwig's Mag. d. Thierheilkunde, iv Jahrg., 2. Heft.
(5) Mayer. Froriep's Neue Notizen i, p. 107 (*T. denticulata*)
(6) ————.Analects f. vergl. Anat. 2. Samml. i, p. 70, figs. 4–5. Review by v. Siebold in Wiegmann's Arch. f. Naturg. 1840, ii, p. 192.
(7) Creplin. Endozoologische Beiträge. 1 Ueber *Taenia denticulata* Rud. und *Taenia expansa* Rud.; Wiegmann's Arch. f. Naturgeschichte, 1842, i, pp. 315–327.
(8) Dujardin. Hist. nat., p. 577.
(9) Diesing. Syst. Helm., i, p. 497.
(10) Verrill, A. E. Parasites of domestic animals; Report of the Connecticut Board of Agriculture for 1870, pp. 65, 100.
(11) Davaine. Traité des Entozoaires, 1877, p. liii, 235.
(12) Moniez. Mémoires sur les Cestodes. Paris, 1881, pp. 15–21, pl. i, figs. 38–58, ii, 1. (Embryology.)
(13) Zürn. Die th. Parasiten, 1882, p. 189–192, taf. iii, figs. 39–41.
(14) Perroncito. I Parassiti, etc., 1882, p. 239, figs. 99–100.
(15) ————Trattato sulle Malattie degli Animali domestici, 1886, pp. 231–232, 233.
(16) McMurrich. Ninth Annual (1883) Report of the Ontario School of Agriculture (1884), pp. 174–178, 7 figs. (Zoological Report.); F. G. Grenside, same Report, pp. 200–203. (Veterinary Report). (Rev. by v. Linstow, Arch. f. Naturg., 1888.) [*T. expansa*=*M. trigonophora*, vide p. 37.]
(17) Leuckart, R. Bandwürmer; Koch's Encyclopädie der gesammten Thierheilkünde, 1885, pp. 396–399; *T. expansa*, under *Dipylidium*.
(18) Railliet. Éléments, etc., 1886, p. 257–260, fig. 151 (orig.), fig. 152 (after Neumann), fig. 153, 1–12 (after Moniez).
(19) Neumann. Traité, etc., 1st ed., 1888, p. 377–378; figs. 164 (1–12) (after Moniez), 165 (after Railliet), 166 (orig.).
(20) Curtice, C.. Animal Parasites of Sheep, 1890, pp. 113–122, pls. xiv, xv.
(21) Zschokke. Recherch sur la structure anatomique et histologique des Cestodes, 1890, pp. 91–114, taf. ii–iii, figs. 31–38.
(22) Blanchard, R. Sur les Helm. des Anthro. (Footnote); Mém. d. l. Soc. Zool. d. France, 1891, p. 187. (*Moniezia expansa*.)
(23) Moniez. Notes sur les Helm., vi, 4; Rev. biol du Nord de la France, 1891.

(24) BLANCHARD, R. Notices Helm., 2 sér., 6; Mém d. l. Soc. Zool de France, 1891, p. 445.

(25) NEUMANN. Traité, etc., 2nd ed., 1892, p. 402, 407; figs. 184-186.—Engl. Transl., 1892, pp. 412. 418, 429. (*T. expansa.*)

(26) DEWITZ. Eingeweidewürmer der Haussäugethiere. Berlin, 1892, p. 77-81, figs. 51-52.

(27) STILES. Notes sur les Parasites—14: Sur le *Tænia expansa* Rud.; Compt. rend. d. l. Soc. d. Biol. (Paris) 1892, p. 664.

(28) ——— Bemerkungen über Parasiten—17: Ueber die topographische Anatomie des Gefässsystems in der Familie *Tæniadæ*; Centralblatt f. Bakterologie und Parasitenkunde, 1893, bd. xiii, p. 457-465, figs. 1, 2. (*M. expansa.*)

HISTORICAL REVIEW.

It is almost impossible to give a satisfactory historical review of this species, since very few of the diagnoses given by authors can be applied to this worm alone.

Rudolphi (1) starts off his discription of the species with the diagnosis—

"*Tænia expansa* R.—Capite obtuso, collo nullo, articulis anticis brevissimis, reliquis subquadratis, foraminibus marginalibus oppositis."

As synonyms and literature he gives:

Bloch.—Abhandlung von der Erzeugung der Eigeweidewürmer und den Mitteln wider dieselben. Berlin, 1782, 4° 10 Taf., p. 16, Taf. 5, figs. 1-5. *Tænia rasis nutritiis distinctis.*

Goeze.—Versuch einer Naturgeschichte der Eingeweidewürmer thierischer Körper. Leipzig, 1782, 44 Taf., p. 360-371, Taf. 28, figs. 1-12. *T. ovina.*

Batsch.—Naturgeschichte der Bandwurmgattung überhaupt und ihrer Arten insbesondere, nach den neuern Beobachtungen in einem systematischen Auszuge. Halle, 1786, p. 182, n. 28, figs. 109, 162. *T. ovina.*

Schrank.—Verzeichniss der bisher hinlänglich bekannten Eingeweidewürmer, nebst einer Abhandlung über ihre Anverwandtschaften. München, 1788, 8°, p. 38, No. 115, *T. ovina.*

Gmelin.—Linné's Systema Naturæ, 1789-1790, p. 3074, No. 55. *T. ovina.* Tabl. Encyl. t. 45, figs. 1-12 (ic. Goeze). *T. ovina.*

Zeder.—Anleitung zur Naturgeschichte der Eingeweidewürmer. Bamberg, 1803, p. 332, n. 7. *Halysis ovina.*

Rudolphi then gives a more detailed description of the animal; but that this description is insufficient is conclusively proven by the fact that many helminthologists since Rudolphi's time have had difficulty in recognizing the proper *T. expansa.* Although Rudolphi acknowledges that his species is identical with the species *T. ovina,* he renames it *T. expansa.* There seems to be an impression among certain authors that the word *expansa* refers to the breadth of the worm, but this is not the case, for Rudolphi writes: "Ipse maximam ab agni pyloro ad cœcum usque expansam et villosæ adhærentem vidi, unde nomen triviale desumsi." Hence Rudolphi referred to the length rather than the breadth in using the term *expansa.*

As far as the literature and synonyms which Rudolphi gives are concerned, the following have been found:

In the French edition of Bloch's work (p. 35-37, v. 1-5) a worm is described which may or may not be considered identical with Rudolphi's

T. expansa. In the former the head is large, and the genital pores, in nearly every segment figured, are exactly in the middle of the margin. The shape of the head is quite similar to that of *expansa*. In *expansa* the head is small, the pores generally in front of the middle of the edge. The worm Goeze describes as *Tænia ovina* could, with a little imagination, be called identical with Rudolphi's *expansa*.

The articles by Batsch and Schrank are not at hand. Gmelin evidently takes his diagnosis from Bloch's article. The other two books are not to be obtained here.

Rudolphi is probably as far back as it is possible to trace *T. expansa* with certainty. Having in my possession part of his type material I feel obliged to accept his name *expansa* in preference to the name *ovina*, which antedates it. If, however, the original material of Goeze can be examined, and if by that means it can be demonstrated that *ovina* is really identical with *expansa*, then I should be in favor of holding to the law of priority and of vindicating Goeze even at this late date. As Rudolphi himself believed his species to be identical with Goeze's *ovina*, he had no excuse for renaming the species.

In his synopsis Rudolphi (2) adds nothing of any importance in regard to *T. expansa*.

Gurlt (3) gives a short description of *T. expansa*, together with two figures. It is probable that this is really *T. expansa*, although it may possibly belong to some other species. It has not been possible to obtain Gurlt's second article.

Mayer (5-6) speaks of *T. denticulata* in two articles (see below).

Creplin (7) gives quite a good discussion of the differences between *T. expansa* and *T. denticulata*. His diagnosis for *T. expansa* reads:

T. capitis parvi oblongi, antice et lateribus rotundati, retrorsum angustati osculis perfecte lateralibus, maguis, tumidis, collo cum corpore toto plano, subtili, cum capite, se crassiore, continuo, perbrevi, articulis primis brevissimis, marginibus lateralibus convexiusculis, sequentibus lentissime latescentibus elongatisque et simul marginibus lateralibus rectioribus demumque fere prorsus rectis, postico recto, sæpe crispato, semper tumidulo, uti angulis obtusis, perparum prominentibus. articulis ultimis solis, cum se antecedentibus sensim angustioribus factis, primo quadratis, tum (perpaucis, neque semper), adeo longioribus quam latioribus, terminali obtuse finito, osculis genitalibus marginalibus plures pollices a capite monstrari incipientibus, tum ad caudam usque, secundis oppositis. (Ova globiformia et angulata.)

Creplin evidently had the true *T. expansa* before him, although it is not impossible that he also had some other species. He says that Mayer's (6) *denticulata* is really *expansa*. He thinks also that Carlisle's specimens were *expansa*.

Dujardin (8) does not add anything new to our knowledge of *T. expansa*. His "large de 5-27 ᵐᵐ; ventouses dirigées en avant" seems to indicate that more than one species is included in the diagnosis.

Diesing (9) gives a short Latin diagnosis of the species.

Verrill (10) evidently has several species included in his diagnosis of *T. expansa.**

Davaine's diagnosis (11) is short. He also gives a short account of symptoms and treatment of animals affected by the parasite.

In Moniez's Mémoires sur les Cestodes we find an account of the embryology of tapeworms of the type *Tænia expansa*. His results are as follows:

The ovum of *T. expansa* first appears as a small cell filled with yolk and provided with a body which seems like an excentric nucleus; this latter, however, is the true ovum. A vitelline membrane appears after fecundation, and the ovum divides; one of the resulting bodies separates from the vitelline matter and multiplies rapidly, forming the blastodermic cells; the other cell remains included in the vitelline matter and becomes very refringent (polar body). The blastodermic cells increase in number, forming a morula, while the vitelline matter contains the polar body. The vitelline matter divides into two semilunar bodies which surround the morula, and each of which contains a large vitreous cell. The morula now gives off in succession two concentric cellular layers; the cells of the outer layer form an envelope for the embryo, then become indistinct and granular. The inner layer of cells then pushes out two horn-like prolongations on one side and forms the so-called pyriform body, inside of which lies the six-hooked embryo (oncosphere). That portion of the outside cellular layer which is included between the two horns of the pyriform apparatus, according to Moniez, forms the prominent circular or radiate body (disk) situated at the end of the horns.

Zürn (13) describes symptoms and gives treatment for animals infested with these tapeworms. His specific diagnosis is short. He goes back to the statement of earlier authors, that *T. expansa* is sometimes 60ᵐ long. His figures are very poor. In Figure 39 part of the segments are upside down, *i.e.*, he mistakes the anterior edge for the posterior edge.

Perroncito's figures (14) are upside down. His diagnosis seems to apply pretty well to *M. expansa*, although some points in it make one suspicious of *M. planissima*. He states, contrary to Creplin, that the suckers are directed forward. The anterior portion of the strobila is described as colorless or light yellowish, while the posterior portion is a more intense yellow. The segments are said to attain a breadth of 33ᵐᵐ.

McMurrich (16) had occasion to investigate a tapeworm epizoötic among sheep in Ontario. He favors a theory that *Melophagus ovinus* (the sheep tick) acts as intermediate host for the tapeworm. Judging from his description and figures it is probable that the worm he found is the same that the writer found later (see below) in the Blairsville epizoötic (*M. trigonophora*).

Leuckart (17) gives a very good general description of this parasite,

---

*In regard to Verrill's report, I might mention that several European authors have concluded that Verrill found in America all the species of parasites which are mentioned in that article. Prof. Verrill, however, has written to this Bureau that such was not the case. As he does not state in every case in his article which of the parasites mentioned are found in America, his report should not be used in making up the geographical distribution of the animals in question.—C. W. S.

but certain portions of his article lead me to believe that he had both *M. expansa* and *M. trigonophora* in his possession when he wrote the article. He discusses the question of life history, and suggests that some snail may possibly be the intermediate host.

Railliet (18) states that the suckers are directed forward, but gives a figure (from Neumann) in which they are directed dorso-and ventro-laterally. An original figure of an adult strobila is given, which seems to be *T. expansa*.

Neumann (19) gives substantially the same as Railliet (18).

Curtice (20) attempted to infect sheep by feeding them eggs of tape-worms. Although his results were not positive, he expresses himself as decidedly in favor of a development without any intermediate host. His experiments cannot be considered as supporting the theory.

Zschokke (21) gives an excellent account of the internal organization of *T. expansa*. There can be no doubt that he had *T. expansa* before him, for the preparations which he kindly forwarded agreed very well with Rudolphi's type specimen. Zschokke's work will be referred to hereafter in connection with the anatomical description.

R. Blanchard (22, 24) placed *T. expansa* in his new genus *Moniezia*.

Moniez (23) discusses the difference between *M. expansa* and *M. denticulata*, stating that the segments of *expansa* are thin, transparent, and yellowish.

Neumann (25) gives a new figure of the head of *T. expansa*. The figure of the head of *T. expansa* given in his first edition is now reproduced as *T. Benedeni*, but it does not agree with Moniez's original material. That Neumann, as well as Railliet, included *M. planissima* in the species *T. expansa* is already stated above.

Dewitz (26) mentions Curtice's conclusions in regard to the direct development and copies one of his figures.

Quite recently I obtained some of Rudolphi's specimens and described (27, 28) the interproglottidal sacs in them; at the same time attention was called to the fact that *M. expansa* of recent authors contained more than one species, and *M. planissima* was separated from *M. expansa*.

Above have been mentioned nearly all the articles upon *T. expansa* Rud. which it has been possible to find. The species is mentioned in several lists of parasites which are not included in this review. It is further to be expected that there are other articles upon a species so well known (by name), but that they have escaped attention.

ANATOMY.

After this tedious review of articles published on *Tænia expansa*, let us pass to a description of that portion of Rudolphi's original specimen in our possession, in order to establish one of the chief points in the diagnosis of this species.

Prof. Möbius very kindly sent four portions of worms labeled "*Tænia expansa;* coll., Rudolphi;" Nos. 1–3 from *Ovis aries,* No. 4 from *Gazella dorcas* (*Antelope dorcas*).

(1) Scolex very much shrunken, nearly square when viewed *en face,* 0.368 by 0.368ᵐᵐ. The suckers occupy the four corners, are distinctly raised and directed diagonally towards the front. The openings are slit-form (Plate v, Figs. 1–1*b*).

(2) A strobila about 10ᶜᵐ long by 1.5ᵐᵐ broad. Poorly preserved. Segments are distinct (Plate v, Fig. 2). Genital organs well developed, but too indistinct to warraut description. Genital pore in anterior half of segment. Interproglottidal sacs almost imperceptible.

(3) Fifteen segments 7ᵐᵐ broad by 1–1.3 long. Interproglottidal sacs plainly visible. The material does not warrant any other statements.

(4) (From *Gazella dorcas.*) Thirty segments; 10 (not 16ᵐᵐ as stated in my Note 14) broad by 1ᵐᵐ long. Somewhat better preserved than 1–3. The segments are filled with eggs; the interproglottidal sacs (Plate v, Fig. 3), are here and there visible.

From the above it will be seen that the species accepted as *Moniezia expansa* must possess interproglottidal glands grouped around sacs. The literature has been searched very carefully to find mention of these glands, but unless some statement by some author has escaped attention, the glands were first described in my preliminary note (26) before the Société de Biologie (Paris).

The question, of course, immediately arises whether these sacs can not be found in other species. As a matter of fact we find these sacs present in several species, which are included in the "*Expansa group.*" By the aid of the scolex, however, we can determine which one of these is the true *M. expansa.* The scolex of Rudolphi's type agrees very well with the scolex of the worm which Curtice described as *Tænia expansa;* it does not, however, agree with Zschokke's figure, although it resembles one of his preparations. The difference between this figure and Neumann's figure might, perhaps, be explained by contraction. At any rate, it is perfectly evident that Curtice's *T. expansa* comes the nearest to Rudolphi's type of any specimen in possession of this Bureau, and justifies the acceptance of that as the true *M. expansa.*

General appearance.—Specimens of this species show an extreme variation in external appearance, according to whether they are alive or fixed by different methods of preservation. These variations, however, are not more than the variations found in specimens of *Tænia saginata* which have been preserved in different fluids. Older authors state that *T. expansa* grows to more than[*] 100 feet in length, but

---

[*] It is well known that several American physicians have reported cases of *Tænia saginata* of man in which the parasites were said to be 50, 100, or even 210 feet long. Zoölogists have explained these cases by saying that the physicians had taken a number of worms, and, supposing them to be all from one individual parasite, have given the total length of all the fragments. I have myself followed this explanation in lecturing upon parasites, and believe that this is the true explanation in many, although not in all cases. Only a short time ago I called upon a physician who has several bottles in his office containing *T. saginata,* and bearing the labels "180 feet long," "185 feet long," "210 feet long," etc. Upon calling into question the supposition

recent authors have already shown that that statement is erroneous, and that this enormous length was made up only, by placing a number of broken strobilæ together and assuming that they all belong to one individual tape-worm. It is not infrequent to find worms 4ᵐ or even 5ᵐ long. The breadth of the worm varies from about 1ᵐᵐ to 16ᵐᵐ, according to the portion (the neck or the posterior end) measured, the age, preservation of the specimen, and the extent of contraction. The anterior portion is generally whitish, while the posterior end is frequently of a yellowish or orange color.

Specimens of worms, agreeing in regard to the shape of the head and presence of the blind sacs, vary considerably in form, but it is not thought best to make separate species for these variations, as it is believed that most of them may be accounted for by contraction.

The *scolex* varies from 0.368–0.72^{mm}; the suckers measure about 0.3 broad by 0.256^{mm} long; the opening is slit-form. The neck is extremely variable, in some cases very thin, in others quite thick (contraction). Segmentation begins from 0.5–2^{mm} back of the head.

Segments 10^{cm} from the head measure 1.5^{mm} wide by 0.24^{mm} long; about 25 sacs are present; the genital anlage is circular in shape, although, in some cases, slight traces of the barrel of the pistol can be seen.

· Segments 15^{cm} back of the head measure 2^{mm} wide by 0.19^{mm} long; about 35 sacs are visible; the genital pores are not yet present; the genital anlage is pistol-shaped, and at the median extremity is already divided into two lobes, the anterior destined to form the male canal, the posterior to form the female canals and glands; the lateral point of the aulage (muzzle of the pistol) is close to the edge of the segment in its anterior third. No testicles are visible.

Segments 20^{cm} from the head (Plate VI, Fig. 3) measure 3.5^{mm} wide by 0.56^{mm} long; about 40 sacs are visible; genital pores not present; the genital canals become more distinct and the ovary, vitellogene and shell-glands can be distinguished; numerous testicles are present; they occupy the median line as well as the rest of the median field. Zschokke states that the testicles of *T. expansa* are arranged in two triangles, one each side of the segment, as is the case in *M. trigonophora* (see below). By far the greater number of segments of *M. expansa* which have been examined do not present this relation, although in a few segments of this species which were collected at Paris,

---

that these bottles contained only one specimen each, I was assured in the most positive terms that the worm "210 feet long" was *entire* at the time of measurement. Further conversation, however, elicited the fact that the worm had not been measured in the way zoölogists take measurement, *i. e.*, by placing the parasite upon a table and measuring its length while lying undisturbed, but that the freshly expelled cestode was in a stool of warm water; one end of it was lifted up and the length taken by means of a yard-stick while the parasite was *in a perpendicular position*. In this way its own weight would, of course, cause an *enormous* extension, and the length of "210 feet" could be easily obtained.—C. W. S.

the triangular arrangement of the testicles is very distinct. This character appears in the same strobila with a quadrangular arrangement, but as most of the segments show the quadrangular arrangement, it is believed the triangular arrangement in the older segments of this species forms an exception. In Rudolphi's specimens the testicles are not visible.

Segments 30^{cm} from the head measure 5^{mm} broad by 0.4^{mm} long; the genital pore has pierced the anterior third of the margin of the segment; the canals are perfectly distinct; the female glands are all visible; the entire remaining portion of the median field is crowded with testicles 56 μ in diameter. (As the segments were subjected to pressure in making the microscopic preparations, this 56 μ is probably larger than normal.)

The topography of the female glands, canals, vagina, cirrus, nerves, and the longitudinal canals agrees with *M. trigonophora* (see below, p. 39–41), so it is useless to repeat the description here. The only difference noticed was that the vagina and cirrus are more exactly on the same transverse plane (Plate VI, Fig. 4). Zschokke states that, as a rule, the vagina is ventral while the cirrus is dorsal, but I have found that the vagina is always ventral on the right side and dorsal on the left side.

The uteri (Plate VI, Fig. 6) come more plainly into view in segments 80–90 ^{cm} from the head, while the genital glands begin to atrophy. As the segments grow wider the interproglottidal sacs increase in number.

In segments 150^{cm} from the head the uteri are fully developed; they extend into the lateral fields dorsally of the longitudinal canals. The testicles are no longer visible; but the receptaculum seminis still persists; this organ is generally somewhat longer than in *M. trigonophora*. In the remaining segments the ova come more plainly into view, and finally the segment represents a bag of ova.

Ova.—The eggs measure 50–60 μ in diameter, the bulb of the pyriform apparatus 20 μ.

SPECIFIC DIAGNOSIS.

The specific diagnosis of this form is as follows:

Moniezia expansa (R., 1810) R. Bl., 1891.—Strobila attains 4–5^m in length; anterior portion usually whitish, posterior portion generally yellowish. Head 0.36–0.7^{mm} in diameter, obtuse, more or less square, slightly lobed; suckers distinctly raised, apertures directed diagonally forward. Segments always much broader than long; end segments attain 16^{mm} in width and are quite thick. Topography of nerves, longitudinal canals, genital canals, and female glands similar to *M. planissima;* testicles usually arranged in a quadrangle, rarely in two triangles except in the younger segments. Interproglottidal glands localized around blind sacs which open between segments. Ova 50–60 μ, bulb of the pyriform apparatus 20 μ.

Type specimen is in the Königl. Museum für Naturkunde, Geheimrath Prof. Dr. Möbius, Direktor, Invalidenstrasse, Berlin, Germany. Several segments of the type are in the Bureau of Animal Industry, and in the private collection of C. W. Stiles. Typical specimens will be sent to museums and specialists exchanging with this Bureau.   C. W. S.

35

(5) Mon|ezia oblongiceps sp. n., 1893.

[Plate VII, Figs. 1-4.]

Host.—*Coassus sp.* (S. & H.)
Geographical distribution.—Washington, D. C., imported from the Island of Trinidad, South America.

In a post-mortem examination of a *Coassus sp?* recently imported for the National Zoölogical Park at Washington, D. C., from South America, two specimens of an adult cestode were found which bear a great resemblance to *M. expansa*, but which differ from it in the form of the head and segments as well as in general appearance.

While for a long time we were inclined to look upon this as a simple variety of *M. expansa*, we incline at present to the view that the parasite in question represents a distinct species, which, however, is very closely allied to *M. expansa* of cattle and sheep. If only a few segments of the worm were in our possession we should undoubtedly determine them as identical with that form, but with two well preserved specimens before us it has been decided that they represent a true species.

ANATOMY.

General appearance.—Two strobilæ provided with heads measured respectively 97^{cm} and 88^{cm} in length. A third strobila, which evidently belonged to one of those provided with a scolex, measured 31^{cm} long. The head is not distinctly separated from the neck; the segments gradually increase in breadth to 50^{cm} from the head, where they reach the maximum of 9^{mm}; from this point they diminish in breadth and increase in length so that the end segments measure 6.5 × 1^{mm}. All the segments are quite thin. When fresh the worms have a very light yellowish tinge.

Scolex.—The head measures (preserved) 0.48–0.56^{mm} broad; it is somewhat narrower in its dorso-ventral diameter. The same measurements are preserved for a short distance back of the head, so that the latter is not distinctly separated from the body. While the head of *M. expansa* viewed *en face* is more or less square, this head is distinctly oblong. The suckers (0.19 in diameter) are situated at the four corners. The openings in our specimens are slit-form and directed half forward, half corner-wise. The suckers are not so distinctly raised as in *M. expansa*.

In Fig. 2b, Plate VII, the portion back of the head is broader than the head; in another specimen (Plate VII, Fig. 2), the diameter of the head is preserved for about 2^{mm}, then it decreases somewhat for a very short distance, and finally increases again. This variation is due entirely to contraction.

Segments 10^{cm} from the head measure 5^{mm} wide by 0.16 long; about 30–36 sacs are present; no testicles are visible, but the genital anlage has the characteristic pistol shape; the pore is not present; the median end of the anlage is already differentiated into a male and female lobe. In segments measuring 5^{mm} broad by 0.48 long, the pore is just on the point of piercing the margin in its anterior third, the outer cuticle invaginating to meet the genital canals; the cirrus-pouch is not yet formed; the male and female canals are separated in their entire length; female glands are all formed but are small. The remaining portion of the median field is occupied by testicles which are found in the median line as well as elsewhere; the interproglottidal sacs have increased in number in proportion to the width of the segment.

Segments 21^{cm} from the head measure 6^{mm} broad by 0.48^{mm} long; the genital pore is present; all of the organs have increased in size; the testicles have increased in number but are less numerous and smaller than those of *M. expansa;* the genital organs and the longitudinal canals present the same topographical relations found in the other species of *Moniezia;* copulation has taken place in segments 22^{cm} from the head. The canal inside the cirrus is frequently quite wide. This widening is sometimes seen in other forms also, but it has never been seen so pronounced as in the worm now under consideration. The genital papillæ are much more prominent than in any other species examined except *M. denticulata* and *Th. Giardi.*

Segments 30^{cm} from the head measure 6.75^{mm} broad by 0.48^{mm} long. Segments 40^{cm} from the head measure 8^{mm} broad by 0.56^{mm} long; the uteri become more prominent, while the female glands begin to atrophy. Segments 50^{cm} from the head reach the maximum breath of 9^{mm}, while in the next 40^{cm} they decrease in breadth to 6.5^{mm} by 1^{mm} long. In the last segments the female glands and testicles have entirely disappeared.

Eggs.—The ova measure 48–60 μ in diameter; the bulb of the pyriform organ 16–20 μ; the horns are very short, 8–10 μ.

SPECIFIC DIAGNOSIS.

The specific diagnosis of this cestode is as follows:

M. oblongiceps sp. n., 1893.—Strobila attains 97 (probably 120 or more) centimeters in length; yellowish. Head not distinctly separated from neck; oblong when viewed *en face,* the breadth (0.48–0.56^{mm}) being greater than the thickness; suckers 0.19 in diameter, not prominently raised, situated at the four corners; openings slit-form and directed diagonally forwards. Segments always broader than long, rather thin; maximum breadth (about 50^{cm} from head) 9^{mm}; end segments 6.5 wide by 1 long. Topography of organs same as in *M. expansa;* testicles less numerous and smaller. Interproglottidal glands arranged around sacs. Eggs 48–60 μ; bulb of pyriform organ 16–20 μ; horns 8–10 μ.

Type in the Bureau of Animal Industry, U. S. Department of Agriculture. S. & H.

(6) Moniezia trigonophora sp. n., 1893.

(Plates VIII, IX.)

Synonymy.—*Tænia expansa* Curtice, *ex parte*, specimens in the collection of the Bureau of Animal Industry. *T. expansa* McMurrich, see p. 30. *T. Benedeni* Neumann, *ex parte.*

Host.—Sheep (Neumann, Curtice, McMurrich, Stiles, Hassall).

Geographical distribution.—North America: U. S. A., Washington, D. C. (Curtice, Stiles, Hassall); Blairsville, Pa. (Stiles); Canada (McMurrich). France (Neumann).

LITERATURE.

McMURRICH. See above, *M. expansa*, p. 27.

STILES. Notes sur les Parasites—14: Sur le *Tænia expansa* Rud.; Compt. rend. d. l. Soc. d. Biol., Paris, 1892, p. 165. (*T. Benedeni* de Neumann.)

——— Bemerkungen über Parasiten—17: Ueber die topographische Anatomie des Gefässsystems in der Familie *Tæniadæ;* Centralblatt f. Bakt. u. Par., 1893, bd. XIII, p. 465.

HISTORICAL REVIEW.

McMurrich (1) mentions a tapeworm epizoötic in Canada, and diagnoses the parasite as *Tænia expansa.* From his figures and description, however, we suspect that the worms belong to this new species.[*]

In September, 1891, Stiles received instructions from Hon. Edwin Willits, Acting Secretary of Agriculture, to proceed to Blairsville, Pa., and study an epizoötic among sheep, which was reported as existing at that place. He found that the sheep were infested with a large number of worms, which were undoubtedly the cause of the trouble. *Strongylus contortus* was present in the fourth stomach in such numbers that it was almost impossible to see the mucous membrane. Besides this, all of the sheep examined contained large numbers (10–50) of tapeworms belonging to the species here described. Upon comparing these worms with parasites in the Bureau collection it was found that Curtice had collected a number of the same kind from sheep slaughtered at Washington, D. C., and had determined them as *Tænia expansa.* As stated above, Neumann's specimen of *M. Benedeni* also agrees with this form.

ANATOMY.

General appearance.—The worms are of a cream color when fresh. The head is quite small; the strobila attains the length of about 1.6 meters. The segments are very sharply defined, in nearly all cases broader than long, although some segments were found which were almost square, or even longer than broad. The largest segments are generally 6–7^{mm} broad by 2^{mm} long.

Scolex.—The head (Plate VIII, Figs. 2–2c) in preserved specimens measures 0.624–0.704^{mm} broad; in a balsam-preparation of the same form

---

[*] Through the kindness of Prof. McMurrich I have been able to examine the worm mentioned by him as *T. expansa*, and have verified the interpretation given above. There is no doubt that the worm is identical with *M. trigonophora.*—c. w. s.

it measured 0.768ᵐᵐ broad. The head passes imperceptibly into the neck, which frequently shows a pseudo-segmentation; the true segmentation begins 2ᵐᵐ back of the head, where the worm is 0.48 broad. The suckers (under pressure) measure 0.272 in diameter. In alcohol specimens the opening is slit-form.

Plate VIII, Fig. 2, represents the head of the smallest worm found. The strobila measured 2.7ᵐᵐ long; segmentation is visible, but no genital organs are developed. The head is pear-shaped and measures 0.48 broad; the suckers (0.256 by 0.2) are directed diagonally forwards—dorso- (or ventro-) laterally.

Segments 50ᵐᵐ from the head measure 2ᵐᵐ wide by 0.13 long; 10–14 interproglottidal sacs are present but indistinct; the genital anlagen are indistinct; no testicles can be seen; 75ᵐᵐ from head, about 18 sacs are visible; no testicles are developed. The genital anlage is represented by a round clump of tissue on each side of every segment.

Segments 10ᶜᵐ (Plate VIII, Fig. 3) from the head measure 0.4ᵐᵐ long by 4.5ᵐᵐ broad; the genital pores are not present; the genital canals are but little differentiated; the median end-knob of the pistol-shaped genital anlage is indistinctly divided into two lobes; the testicles and ovaries are not visible. At the border between every two segments a number (20–27) of blind sacs are distinguishable; 150ᵐᵐ from head the segments are 3.5ᵐᵐ wide by 0.4ᵐᵐ long; testicles begin to develop; the genital canals gradually become more distinct.

In segments 190ᵐᵐ from the head the testicles are numerous. In segments 200ᵐᵐ from the head, the female anlage becomes quite distinct from the male ducts. It sends a prolongation toward the center of the segment; a portion branches from this toward the posterior edge of the segment and becomes the anlage of the vitellogene gland; another portion extends toward the anterior edge; its end spreads and forms the anlage of the ovary. Between these two anlagen is seen the anlage of the shell gland. The portion of the male anlage just median of the longitudinal canals begins to form convolutions, and its end, which is rather thin, can be traced past the female glands a short distance toward the median line of the segment. At about this stage the canals, which were at first solid, begin to show a lumen.

Segments 30ᶜᵐ (Plate VIII, Fig. 4) from the head measure (balsam preparation) 6ᵐᵐ broad by 0.88 long. The interproglottidal sacs have increased in number (30–36) and size. The genital pores have pierced the cuticle; they are invariably situated in the anterior half of the margin. The cirrus-pouch is well developed; the lateral portion of the vas deferens (*vesicula seminalis*) is much convoluted and filled with spermatozoa; the median portion of the vas deferens crosses the ovaries; 150–240 testicles are present in each segment and are arranged in two triangles; the long leg (base) of the triangle is nearly parallel to the posterior edge of the segment, the short leg (perpendicular) parallel with the lateral margin; the hypothenuse runs from the antero-lateral portion to the posterior edge near the median line. While the testicles are generally included within these limits, they are frequently found outside the same, so that the triangles must be considered as a convenient but not rigid diagram.

Occasionally testicles were found in the lateral fields. This shows an initial step in the wandering of the testicles from the median to the lateral fields, a tendency which reaches its highest development in *Th. Giardi*. In the female genital organs, the canals present distinct lumens; the vagina is sharply separated from the receptaculum seminis; the three glands are distinctly differentiated from each other, both anatomically and histologically.

Segments 60^{cm} (Plate IX, Fig. 4) from the head measure (under pressure) 6^{mm} broad by 1^{mm} long. There are 35-37 interproglottidal sacs to each segment. Copulation has already taken place, as is shown by the presence of spermatozoa in the receptaculum seminis. The various genital organs, with the exception of the uteri, can all be distinguished. The female glands form a circular clump plainly visible to the naked eye on stained specimens. For a careful study of the anatomy, segments 60-80^{cm} from the head are well adapted. The segments are nearly the same width (6-6.5^{mm}), but increase gradually in length (1-1.36^{mm}). The posterior flap of every segment overlaps the next segment by 0.24^{mm}. The arrangement of the testicles in two triangles is general, but testicles are frequently seen in the median line. The arrangement of the female genital glands in a rosette is very striking. In the general topographical relations of the organs there are noticed several important characteristics which agree with *M. planissima*, i. e., the genital pores are in the anterior half of the margin; on the right side the vulva is ventral of the cirrus-pouch, and on the left side the cirrus-pouch is ventral of the vulva; the two organs are, however, not exactly on the same transverse plane, but the vulva is regularly found slightly posterior to the cirrus. The genital canals of each side, in passing to the median field, cross the two longitudinal canals and nerve dorsally, and the vas deferens passes the ovary dorsally. The relative position of the genital and longitudinal canals, or of the ovary and vas deferens, gives us two excellent points by which we can immediately distinguish the dorsal from the ventral surface.

Male genitalia.—The description of the cirrus-pouch of *M. planissima* applies almost equally well to this species, except that the layer of ciliary projections is not well developed; in fact, in many cases it was impossible to find it. As a general rule, a genital cloaca is present instead of a genital papilla, even when the penis is extended. That, however, is simply a matter of contraction, for, as seen below (*Th. Giardi*), the development of the papilla in its fullest extent depends upon the absence of the cloaca and vice versa. In the genus *Moniezia*, however, there is a distinct collar around the genital openings even when the genital cloaca is reduced to a minimum. From the base of the cirrus-pouch to the free end of the extended penis is 0.32^{mm}; to the base of the cloaca, 0.16. The median half of the pouch is considerably constricted, owing to the presence and contraction of a thick layer of circular muscles. The vas deferens branches freely, the branches running to the testicles, which are about 48 μ in diameter.

Female genitalia.—The only practical difference observed between the vulva of this species and that of *M. planissima* is that the former has a position slightly posterior to the center plane of the cirrus-pouch. The vagina is about 0.4^{mm} long; the histology agrees with that of *M. planissima*, except that the muscular layer is slightly more developed. The ciliary layer is somewhat less developed; in fact, in most cases, it has not been possible to recognize it at all.

The receptaculum seminis lies dorsally of the ovary, but generally presents a slightly different aspect from that of *M. planissima* or of *M. expansa;* while in those latter species it was rather long and not very wide, in this species it is much shorter in proportion to its breadth $(0.28^{mm}$ long by 0.144). The wall consists of a simple layer of flat epithelium; in many cases it has been impossible to distinguish any basement membrane, while in several cases it was distinctly visible. As a general rule it may be stated that the receptaculum seminis runs diagonally from before backwards instead of parallel with the anterior edge; this, however, may be subject to variation according to the state of contraction of the segment.

From the ventral side of the receptaculum seminis (Plate VIII, Fig. 5; Plate IX, Fig. 3) near its median extremity a canal takes its origin and makes a bend ventrally. In this canal the ova come in contact with the spermatozoa (fertilization canal or first portion of the oviduct). The canal can be traced to the calyx of the ovary, with which it is connected by a constricted portion, as in the case of *M. planissima.*

A short distance from the receptaculum seminis this oviduct gives off a canal dorsally, which, after one or two convolutions, enters the shell-gland and runs to its center. Here it meets with the vitello-duct, and from their point of juncture (ootyp) the third portion of the oviduct extends towards the uterus after making several convolutions in and outside of the shell-gland. As it passes towards the anterior portion of the segment it runs directly dorsal of the upper end of the first portions of the oviduct and empties quite suddenly into a cavity elongated laterally, which is the uterus. The exact positions of these canals vary slightly in different segments.

Histology.—The histological structure of all these canals agrees very well with that described in *M. planissima;* i. e., there is a lining epithelium with basement membrane; the constriction at the upper extremity of the oviduct is surrounded by circular fibers. The ova follow the same course as those in *M. planissima;* i. e., from the calyx of the ovarium they descend through the constriction to the first portion of the oviduct (fertilization canal), come in contact with the spermatozoa, descend through the second division of the oviduct to the middle (ootyp) of the shell-gland, coming in contact with the vitelline matter and egg-shell secretion; they then pass through the ascending (or third) portion of the oviduct to the uterus.

Excretory canals.—The ventral canals are much larger than the dorsal canals, and are connected at the posterior border of each segment by the transverse canal. As a rule they lie laterally and ventrally of the dorsal canal; rarely a projection of the ventral canal extends somewhat dorsally of the dorsal canal (probably a matter of contraction). The lumen of the ventral canals measures 0.368^{mm} by 0.064^{mm}, being longer laterally than dorso-ventrally. It is lined by an extremely thin cuticle, surrounded by parenchyma. The muscle-fibers of the body cluster around it so that it sometimes appears as if they formed a regular layer around the lumen.

The dorsal canal seldom measures over 24μ by 9μ; its cuticle is thicker than that of the ventral canal. Sudden swellings are seen, such as are described below in *Th. Giardi.*

Interproglottidal glands.—The cells of the interproglottidal glands are grouped, as stated above. Sections through these (Plate IX, Figs. 1–2) show that numerous large cells are grouped around small blind sacs, which empty between the proglottids. The entire structure is about 0.15^{mm} in diameter. The nuclei of the cells measure 4μ. While the nuclei of these cells in *M. planissima* generally contained only one nucleolus, two nucleoli are observed to be the rule in this species; the cells converge towards the sac. The wall of the sac has the same histological structure as the vagina and cirrus, *i. e.*, the lumen is surrounded by a layer of ciliary projections (5μ), which is followed by a nearly homogenous layer (3μ) containing nuclei. It was seen above that the vagina was not an invagination of the outer cuticular wall, and the similarity of histology between the vagina and these sacs would naturally lead one to assume that these sacs also are not invaginations in the strictest sense of the term.

Segments 100^{cm} from the head measured, under pressure, 6.5^{mm} broad by 1.12^{mm} long (median line), 1.28 (on the side). The blind sacs have increased to about 40 in number. In segments 100^{cm}–150^{cm} from the head the breadth varies from 5.5 to 6.5^{mm}, the length from 1.25-2^{mm}. A decided change is brought about in the anatomy of the segment by the change in the genital organs. The testicles, ovaries, vitellogene, and shell-glands gradually atrophy. The two uteri (Plate IX, Figs. 5, 6) come more plainly into view, one on each side of the segment; they form numerous folds and gradually occupy the lateral and median fields, generally leaving, however, a clear space in the median line. The receptaculum seminis is full of spermatozoa and very prominent.

In connection with the above description it must be mentioned that the point (*i. e.*, the distance from the head) where the various organs begin to develop varies greatly in different specimens. In some cases the interproglottidal sacs appear before the genital organs, in other cases not until after. In several specimens of this species the sacs were very few or even entirely absent (young specimens), while in one specimen they were fused into a regular line very similar to the interproglottidal line of *M. planissima*. This is undoubtedly to be looked upon as a reversion.

In the end segment of one specimen was found a single median but normally developed set of female genital organs with a single lateral pore. This probably represents a reversion to a former type with single pores.

The mature ova measure 52 μ—60 μ in diameter; the bulb of the pyriform apparatus is 20 μ-24 μ; the horns (12 μ-15 μ long) end in a knob; the hooks of the embryo measure 9 μ long. In this species it was plainly seen that the pyriform body was surrounded by a membrane, a detail which was not so clear in *M. planissima*.

SPECIFIC DIAGNOSIS.

The specific diagnosis of this form is as follows:

M. trigonophora sp. n., 1893.—Strobila attains 1.6-2^m in length, cream to whitish in color. Head 0.624^{mm}-0.704^{mm} broad; suckers not distinctly raised, 0.256 by 0.2^{mm}, slit form. Neck filiform, 2^{mm} long. Segments generally broader than long, although end segments are occasionally seen which are square or even slightly longer than broad; rarely over 6^{mm} broad by 2^{mm} long. Topography of longitudinal canals, genital canals, and female organs the same as in *M. planissima* and *M. expansa;* testicles usually arranged in two triangles; generally absent from the median line of segment. Genital pore never behind the middle of the segment. Ova 52-60 μ; bulb of pyriform apparatus 20-24 μ; horns 12-15 μ.

'Type in the Bureau of Animal Industry, U. S. Department of Agriculture. Typical specimens will be sent to museums and specialists exchanging with this Bureau. S. & H.

C. Denticulata Group. (*M. denticulata*, and *M. alba*.)

Interproglottidal glands absent.

(7) **Moniezia denticulata** (R., 1810) R. Bl., 1891.

[Plate V, Figs. 4-7.]

Synonymy.—*Taenia denticulata* Rudolphi, 1810; *T. denticolata* (R.) Perr., 1882; *Moniezia denticulata* (R) R. Bl., 1891; (*Alyselminthus denticulatus*, Blainville, after Baird, 1853).

Host.—Cattle (Rudolphi).

Geographical distribution.—? (Alfort Museum, Rnd.'s specimens.)

LITERATURE.

(1) RUDOLPHI. Hist. nat. ii, 2, pp. 79-81.
(2) ——— Ent. Syn., p. 145.
(3) GURLT. Lehrbuch der pathologischen Anatomie der Haussäugethiere, 1831, i, p. 381, taf. x, figs. 3-4.

(4) CREPLIN. Endozoologische Beiträge. 1. Ueber *Tænia denticulata* Rud. und *Tænia expansa* Rud.; Wiegemann's Archiv. f. Naturg., 1842, i, pp. 315–327.
(5) DUJARDIN. Histoire nat., p. 578.
(6) DIESING. Syst. Helm., i, p. 498.
(7) BAILLET. Histoire naturelle des Helminthes des principaux mammifères domestiques, 1866, pp. 162–163.
(8) DAVAINE. Traité, etc., 1877, p. liii.
(9) ZÜRN. Die tierischen Parasiten, p. 197.
(10) PERRONCITO. I Parassiti, etc., 1882, p. 240, figs. 101–102.
(11) LEUCKART, R. Bandwürmer; Koch's Encyclopädie der gesammten Thierheilkunde, 1885, pp. 399–400. *T. denticulata* under *Dipylidium*.
(12) PERRONCITO. Trattato, etc., 1886, p. 232, fig. 78 (after Gurlt), 80–81 (orig.).
(13) RAILLIET. Éléments, etc., 1886, p. 260.
(14) NEUMANN. Traité, etc., 1888, 1 éd., p. 377.
(15) R. BLANCHARD. Sur les Helm. des Anthr. (foot note); Mém. d. l. Soc. Zool. d. France, 1891, p. 187.
(16) MONIEZ. Notes sur les Helminthes, VI, 4; Rev. Biol. du Nord de la France, 1891, 3 pages.
(17) R. BLANCHARD. Notices Helmin. 2. sér., 6; Mém. d. l. Soc. Zool., 1891, p. 444.
(18) NEUMANN. Traité, etc., 2 éd., 1892, p. 401, Fig. 183. Engl. Transl., pp. 412, 429; Fig., 183 (*M. Benedeni*).
(19) CURTICE, C. Parasites, etc., 1892.
(20) Stiles. Bemerkungen über Parasiten—17: Ueber die topographische Anatomie des Gefässsystems in der Familie *Tænaidæ;* C. f. B. u. P., 1893, bd. XIII, p. 465.

HISTORICAL REVIEW (1810–1892).

Rudolphi gives as diagnosis of *Tænia denticulata:*

Tænia: capite tetragono, collo nullo, articulis brevissimis, foraminibus marginalibus oppositis, lemniscis dentiformibus.

Camper. in Beschäf. d. Berlin. naturf. Freunde, IV, p. 39.
Gmelin. Syst. Nat., p. 3074, n. 55. *Tænia ovina, β bovis.*
Carlisle. Trans. Soc., Linn. II, tab. 25, Fig. 15–16. *T. ovina, boris.*
Rudolphi. Hodœpor vol. I, p. 81; vol. II, p. 39. *T. denticulata.*
Hab. in Bove. Camperus a bove; Havemannus, Scholæ Vet. Hannov. Director merittissimus, a vitulo (solitariam, capite destitutam); Chabertus a vacca copiose, dejectam observaverunt, hic etiam in vaccæ ventriculo quarto reperit. Specimina mea ex Museo Scholæ Veterinariæ Altorfensis ditissimo.

Descr. Vermes quindecim ad sedecim pollices longi, antice duas ad quinque lineas, postice fere pollicem lati; coloris albidi, vel grisei.

Caput exiguum, tetragonum, latiusculum, *osculis* quatuor, anticis, subcontiguis, subglobosis, horum apertura exigua, orbiculari. *Collum* nullum. *Articuli* aliquot capiti proximi angusti, mox vero latiores, tandem latissimi et subæquales fiunt, ita tamen, ut in uno alterove specimine media vermis pars paullo angustiores et simul longiores objiciat; omnes ceteroquin brevissimi, ut longitudo latissimorum vix lineam, plurimorum ne dimidiam quidem excedat. Margines articulorum postici crenati sive undulati superficiem insequentium qua partem tegunt; laterales (anterioribus nonnullis, rarius medius quibusdam articulis exceptis) obtusiusculi, foramine utrinque medio, opposito, insignes, e quo denticulus acutus, leviter reflexus et duriusculus (lemniscus) exseritur. Substantia mollis, plus minus crassiuscula, ut crassities interdum lineam adæquet. Ova in substantia media cumulata, ovariis regularibus mihi non visis.

Of the articles which Rudolphi mentions, Camper has not been consulted. It does not seem that Carlisle's figures could refer to this species, and it seems doubtful whether Gmelin's reference is to *denticulata*.

Gurlt (3) obtained Rudolphi's specimen and gave two figures, which are copied (Plate v, Figs. 4, 5).

Creplin (4) does not believe that Carlisle's drawings refer to *T. denticulata*. Mayer's drawings (Analecten f. verg. Anat., 2. Samml., Figs. 4, 5) he thinks are *T. expansa*, and he is skeptical whether Camper's and Havemann's specimens were *T. denticulata*. Creplin himself obtained Rudolphi's original specimens, and says in regard to them that there was only one specimen provided with a head; this specimen measured 1–2″ long. The head was rather four-cornered, and sat like a knob on the end of the anterior cone-shaped portion; the suckers formed the blunt corners of the head and opened directly in front (Plate v, Fig. 5). The portion immediately following the head was wrinkled, but it was not clear whether these wrinkles represented true segments, hence it was doubtful whether a neck was present or not; this portion widened very rapidly. All the segments were very short; the lateral edges were convex; the posterior edge covered the following segment for some distance. The broadest segments he had were about ½″ wide by 1‴ long, thick and bloated. Double pores were present. *These were situated immediately behind the posterior edge of the next anterior segment, with the exception of a few cases where they were nearer the middle;* [*] from the pores the lemniscus (cirrus) protruded; the latter was short, thick, and rather conical, on the end truncate.

Creplin then describes the ova with the pyriform apparatus. As diagnosis of *T. denticulata* he gives the following:

T. capitis tetragoni, parvi osculis angulos ipsius efficientibus, magnis, antrorsum apertis, collo subnullo, articulis omnibus crassis, celeriter insigniterque latitudine, paulum longitudine crescentibus, anterioribus ideo brevissimis, reliquis omnibus perbrevibus atque admodum latis, marginibus horum lateralibus convexis, postice protracto et ita sequentis articuli partem anteriorem circumcirca late tegente, foraminibus articulorum latiorem (et lemniscis) marginalibus oppositis secundis. (Articuli postici incogniti. Ova globiformia.)

Dujardin (5), Diesing (6), Davaine (8), and Zürn (9) do not add any original observations on *T. denticulata*, but give very short diagnoses.

Baillet (7) states that *T. denticulata* measures 35 to 40 or even 78ᶜᵐ in length. He gives the measurement of the eggs as 0.09–0.095ᵐᵐ, which, as Moniez has already pointed out, is an error. In his description there is little which aids us in determining this species.

Perroncito (10) considers *denticulata* rather as a variety of *expansa* than as a separate species, but it is evident from his figures and description that he did not have a true *denticulata* before him. His figures seem to me to resemble *expansa*, while a part of his description applies equally well to *expansa* and *planissima*.

---

[*] The italics are mine. From my drawing of Rudolphi's type (Fig. 6) it is evident that Creplin made his statement on loose segments, and mistook the anterior for the posterior edge.—C. W. S.

Lenckart (11) has published by far the best description of this species that has yet appeared (Creplin excepted). It is based upon a specimen in the Leipsic museum. Leuckart states:

Eine zweite, dem letztern eigenthümliche Art ist die gleichfalls den Dipylidien zugehörige *Tænia denticulata* Rud. In ihrer Statur der *Tænia expansa* ähnlich, unterscheidet sie sich aber nicht bloss durch eine ungleich geringere Grösse, sondern schon auf den ersten Blick, gewisser weiterer Eigenthümlichkeiten einstweilen nicht zu gedenken, durch ein gedrungeneres, derberes und festeres Aussehen. Ein Exemplar von 320^{mm} Länge hat schon 25^{mm} hinter dem Kopfende eine Breite von 4^{mm} und in dem 100^{mm} langen hinteren Körper eine solche von 13^{mm}. Die Länge der letzten Glieder beträgt 1.5^{mm}. Bei grösseren Würmern soll die Breite sogar auf 26^{mm} heranwachsen. Der Kopf ist etwas grösser als bei *Tænia expansa* (0.7^{mm}) und mit grösseren Saugnäpfen versehen, die ihm eine mehr viereckige Form geben. Auf ihn folgt auch hier zunächst ein dünnerer Abschnitt, der aber sehr bald schon sich verbreitert und gliedert, und weit rascher als bei *Tænia expansa* zu den späteren Dimensionen heranwächst. Dabei sind die Glieder aber durchschnittlich kürzer als bei der letztern. Die auch hier mit zunehmender Grösse allmählig immer stärker hervortretenden manschettenförmigen Gliedränder sind gewöhnlich deutlich gewellt und auch über die Seitenkanten hinaus fortgesetzt, so dass die hinteren Ecken stärker prominiren und an den letzten und grössten Gliedern zahnartig nach aussen vorspringen. Die reifen Eier gleichen denen von *Tænia expansa*, bleiben aber mitsammt dem Embryonalkörper in ihren Dimensionen um Einiges hinter denselben zurück.

Leuckart seems to think that this species is not an uncommon parasite, although, as he has recently written (personal correspondence), this opinion is based upon report more than upon his own observation.

Railliet (13) states that the eggs are cuboid and twice as large as those of *T. expansa*.

Neumann adds nothing new to our knowledge of this parasite in the first edition of his Traité (14).

R. Blanchard (15, 17) places *denticulata* in his new genus *Moniezia*.

Moniez (16) states that *denticulata* is frequent in sheep killed at Lille. He gives as characteristic of this parasite that the segments are opaque and thick, and the square head is different in shape from that of *expansa*. Moniez also cites Baillet's paper, and discusses Creplin's article.

I am at a loss to know how to judge this article by my distinguished colleague, for, although his statements are concise, it was found upon comparing specimens of the parasite he determined as *T. denticulata* that they did not agree with Rudolphi's specimens. Hence I am compelled to assume, for the present at least, that Moniez has not examined the true *M. denticulata*.

Neumann (18) lengthens his diagnosis in the second edition of his excellent work, and gives a figure of the scolex of *T. denticulata*. This figure, however, does not agree with Gurlt's figure of Rudolphi's type, but agrees exceedingly well with my drawing of Moniez's original specimen of *T. Benedeni*. Hence it must be assumed that Neumann also did not find the true *M. denticulata*.

Neumann quotes Humbold (Berliner thierärztlich. Wochenschrift,

1892) as stating that he found a specimen 45 meters long in a five-year-old cow. Humbold's statement must, of course, be taken *cum grano salis*. Curtice (19) is the only author who claims to have found this species in America. His specimens, however, which are now in our Bureau collection, are certainly not *M. denticulata*, although they are too macerated to be definitely determined.

From the above review it must be concluded that scarcely anything is known about *M. denticulata;* that the only authentic specimens of this species which have found their way into print are Rudolphi's specimens which he received from Alfort, and the specimen Leuckart described. We know practically nothing of the microscopic anatomy of *M. denticulata*.

Through the kindness of Geheimrath Karl Möbius, the writer has in his possession a few segments of Rudolphi's original *Tænia denticulata*. These segments are filled with ova; they measure 8^{mm} broad by 1.14^{mm} long. The specimens are so old that it would be unsafe to state much in regard to them, but several points are distinctly visible. In the first place, one is immediately struck with the position of the cirrus, which pierces the side of the segment (Plate v, Fig. 6) at or very near its posterior border. The cirrus is stout and blunt; the cirrus-pouch is 80 μ long, the proximal portion (24 μ) being constricted by circular muscles, as observed in other species. The vas deferens could be traced but a short distance; it could, however, be plainly seen that the genital canals cross over into the median field dorsally of the longitudinal canals. It will be noticed also that the lateral field is very narrow, the ventral canal coming very close to, or even crossing, the proximal portion of the cirrus-pouch. The outer canal can be distinctly seen to occupy a position slightly ventrad of the other canal, and to have a thinner cuticle. Hence the former, by analogy, is the ventral, the latter the dorsal canal. The transverse canals were not visible. The genital glands have all been suppressed by the uteri. The uteri were confined to the median field. The ova are round (48 μ in diameter), the embryo measures 12 μ in diameter and is surrounded by a pyriform apparatus. No interproglottidal glands could be discovered.

With the aid of the three figures of this species given in the present paper, it is hoped that some one will be able to determine *Moniezia denticulata* and give an account of its microscopic anatomy.

SPECIFIC DIAGNOSIS.

The following may be accepted as a provisional diagnosis of this species:

Moniezia denticulata (R., 1810) R. Bl., 1891.—Strobila $32-40^{cm}$ (to 78^{cm}) long; head 0.7^{mm} broad, with proportionately large suckers directed anteriorly; neck absent or very short, anterior segments very short, increasing rapidly in breadth, but gradually in length as they grow older (Gurlt's figure); 25^{mm} from head they measure 4^{mm} wide;

segments may attain 13ᵐᵐ to nearly 25ᵐᵐ (R.) in breadth; mature segments quite thick; segments 8ᵐᵐ broad by 1.14 long are filled with ova. Genital pores are in the posterior half of the margin; in segments of Rudolphi's type they are close to the posterior corner of the segment (cf. also Gurlt's figure); penis prominent and blunt; relation of vas deferens (and probably of the vagina) to the longitudinal canals the same as in *M. planissima;* uteri (always?) confined to the median field; eggs 48 μ, bulb of pyriform body 12–16 μ, horns 8–16 μ.

Type in the *Konigl. Museum für Naturkunde*, Berlin. A few segments of the type are in the private collection of C. W. Stiles.

C. W. S.

(8) Moniezia alba (Perroncito, 1879) R. Bl., 1891.

[Plate X.]

Synonymy.—*Tænia alba* Perroncito, 1879; *Moniezia alba* (Perr.) R. Blanchard, 1891; *M. alba* var. *dubia* Moniez, 1891.

Hosts.—Sheep (Perr., Mattozzi, Neumann, Blaise); cattle (Perr., Moniez, Railliet).

Geographical distribution.—Italy (Turin, by Perroncito; Macerata, prov. Marches, by Mattozzi). France (Lille, by Moniez; Alfort, by Railliet). Algeria (by Blaise, after Neumann).

LITERATURE.

(1) PERRONCITO. Di una nuova specie di Tænia (*T. alba*); Analli della R. Accad. d' Agric. d. Torino, 1879.

(2) ———. Ueber eine neue Bandwurmart (*Tænia alba*); Arch. f. Naturgeschichte, 1879, 45. Jahrg., I, pp. 235–237; taf. XVI, figs. 1–10.

(3) MONIEZ. Note sur deux espèces nouvelles de tænias inermes, *T. Vogti* et *T. Benedeni;* Bull. scient. d. Dept. d. Nord (1879) 1830, p. 163.

(4) PERRONCITO. I parassiti, etc. 1882, pp. 243, 244, figs. 103–105.

(5) ZÜRN. Die tierischen Parasiten. 1882, p. 197.

(6) LEUCKART. Bandwürmer; Koch's Encyclopädie der gesammten Thierheilkunde, 1885; *T. alba*, p. 400.

(7) RAILLIET. Éléments, etc. 1886, p. 260.

(8) PERRONCITO. Trattato, etc. 1886, pp. 233–234, figs. 79, 82.

(9) NEUMANN. Traité, etc. 1. éd., 1886, pp. 378, 382.

(10) ———. Observations sur les ténias de moutons; Compt. rend. d. l. Soc. d' Hist. Nat. d. Toulouse, 1891, 18 mars.

(11) BLANCHARD, R. Sur les Helminthes des primates anthropoides; Mém. d. l. Soc. Zool. d. France, 1891, p. 187 (foot note).

(12) MONIEZ. Notes sur les Helminthes, VI, 5; Rev. Biol. d. Nord d. l. France, 1891, IV.

(13) BLANCHARD, R. Notices Helminthologiques. 2 sér., 7; Mém. d. l. Soc. Zool. d. France, 1891, p. 444.

(14) NEUMANN. Traité, etc. 2. éd., 1892, pp. 403–408, Fig. 187; Engl. Transl., pp. 412, 418.

(15) STILES. Notes sur les parasites—14: Sur le *Tænia expansa* Rud; Compt. rend. d. l. Soc. d. Biol., Paris, 1892, p. 665.

HISTORICAL REVIEW.

Perroncito's Italian article (1) is not at my disposal, but he printed an article (2) in Germany about the same time, and that probably contains substantially the same data. The description reads:

T. candida, multo rarius hinc inde tractu quodam diluto-flavescente vel flavido-terreo (ochreo) intertincta, sat procera, eximie elastica, metr. 0.60–2.50 longa. Statu recente (mox ex intestino sublata) pallida apparet articulisque s. proglottidibus elongatis, saepe longioribus quam latis, efformata videtur—qui articuli deinde contrahuntur magisque lati quam longi fiunt.

Caput subgloboso-quadrangulare, sat distinctum, mm. 1.40–1.15 latum, vix ultra mm. 1 longum, acetabulis (s. osculis suctoriis) praeditum orbicularibus vel subovalibus, extrorsum ac sursum spectantibus, diametro mm. 0.356–0.456.

Collum breve, saepe depressione circulari constrictum, mm. 1.5–5.320 longum, mm. 0.6–0.912 crassum.

Proglottides, quarum summae mm. 0.020–0.038 longae atque valde angustae sunt, sensim longiores et latiores fiunt, formamque assequuntur rhomboideam, subcampanulaceam, angulis posticis obtusis, ita prominentibus, ut partim proglottidem subsequentem obtegant corpusque taeniae sat acute serratum appareat.—Quae e. dem 1 a capite distant, jam ultra mm. 3 longae et c. tertiam mm. partem latae sunt.—Quae medium taeniae tractum efficient, longitudinem mm. 3–3.5, latitudinem 4–5 mm. praebent atque poris genitalibus manifestis, in quavis proglottide binis, oppositis, inter tertiam proglottidis partem anteriorem et mediam sitis, instructae sunt. Cirrhus plerumque prominens ac tenuis, conicus vel cylindraceus, rectus aut varie incurvus (saepius retrorsum deflexus).—Pars taeniae postica crassior, articulis brevioribus ac latioribus, circa 1.5 mm. crassis, mm. 2–3.5 usque ad 4–5, raro 5-6-6.5 longis, mm. 8.5–9.5, raro ultra 10, rarissime 12–14 longis efformata. Proglottides maturae plerumque contractae ceruntur ac valde breves (2–3 mm. longae), uteris binis (utrinque singulo) instructae, qui loculis creberrimis, invicem coniunctis, constant, unde, proglottide puncta, ingens ovulorum copia prorumpit.

Ovula candida, cubica (quavis facie quadrata vel rectangulari, angulis rotundatis, lateribus convexis, mm. 0.048–0.052, subinde etiam 0.058 longis), testam praebent duriusculam, chitino efformatam atque guttulis adiposis variae magnitudinis obtectam. Matura embryonem fovent, qui corpore sphaerico at duobus processibus aucto constat ac plerumque situm diametralem in ovulo tenet (corpus enim embryonis plerumque angulo huic aut illi adiacet, dum processus oppositum ovuli angulum petunt). Corpus, scilicet pars embryonis princeps mainscula, sphaerica et homogenea, utriculum constituit, cuius media protoscolex inclusus est globosus, protoplasmate homogeneo (dense at pallide granuloso) nucleisque 2–4–5 efformatus atque unculis longiusculis 6 praeditus, quorum ope se in utriculo movet rotatque. Processus supra dicti ab illo corporis latere, quod centrum oculi spectat (exiguo inter ipsorum exortum spatio interjecto) prodeunt, inter se convergentes ac denique aut decussati aut valde appropinquati: alter glochidiatus, scilicet pedicello constat gracili sub-conico, sensim attenuato, apice in cucullum sive appendiculam semilunarem (antrorsum convexam, retrorsum concavam, ad latera utrinque acuminatam ac recurvam) expanso; alter praecedentis pedicello consimilis et aequilongus, exappendiculatus, illius appendiculum denique attingit (cum ipsa tamen non confluens).

The author also adds a short differential diagnosis between *alba*, *expansa*, and *denticulata*. The ten figures are good so far as outline goes, but details are lacking. The parasites were found both in cattle and sheep. By comparing this with Perroncito's next article it is noticed that in giving the measurements above " 12–14 *longis efformata*," *longis* is a misprint. The breadth is here referred to.

Moniez (3) states that he has found this species quite common at Lille, France.

Perroncito (4, 8) gives substantially the same diagnosis in his work on parasites which he gave before (2). Zürn (5) copies his diagnosis from Perroncito's Italian article. Railliet (7) states that he finds very little difference between this species and *T. expansa*. Neumann (10) states in his analytical table that the segments become longer than broad, and do not exceed 10^{mm} in width. Blanchard (11, 13) places *T. alba* in his new genus *Moniezia*. Moniez (12) describes a *M. alba* var. *dubia* which, he states, is intermediate between *T. alba* and *T. expansa*.

Prof. Perroncito has kindly sent to me three specimens of his original type material from which he described this species, and Prof. Neumann has sent a specimen of a worm which he diagnosed as *T. alba*. Unfortunately it has been almost impossible to stain Perroncito's specimens satisfactorily, hence it is hazardous to make many definite statements in regard to them.

ANATOMY.

General appearance.—In regard to the anatomy of the segments, there is no doubt that the topographical relations of the vulva, cirrus, vagina, vas deferens, ovary, shell gland, vitellogene gland, uterus, longitudinal nerves, dorsal, ventral, and transverse canals are the same as found in the other species of *Monieza*. The testicles are arranged in a quadrangle. No interproglottidal glands of any description could be distinguished. Had sacs been present they would undoubtedly have been discovered, notwithstanding the poor staining, for the heavy cuticular lining of the sacs always stands out prominently, no matter how macerated the specimens are. As the head and segments are totally different from *M. expansa*, and as no sacs are present, it can therefore be stated with certainty that *M. alba* is not identical with *M. expansa*. In some segments of other species, which were even more macerated than Perroncito's three specimens, it has been possible to distinguish linear interproglottidal glands. As they were not seen in Perroncito's specimens, it is assumed that no interproglottidal glands are present in the species *M. alba*.

Ova.—The eggs measure 64–68 μ; the bulb of the pyriform body 20–24 μ; the horns 12–20 μ.

Prof. Neumann's specimen is in somewhat better condition than the type specimens, and is evidently identical with Perroncito's *Tænia alba*. The total length is 105^{cm}. The head is too distorted to warrant description; it does not, however, bear any very great resemblance to the head figured by Neumann as *T. alba*.

Segments 100^{mm} from the head measure 2^{mm} wide by 0.4 long. The median field is filled with testicles, which are as numerous in the median line as elsewhere. The genital canals and female glands have the characteristic pistol-shape, the muzzle of the pistol being situated near the lateral edge of the anterior third of the segment. The anlagen of male and female canals are distinct; the female glands represent a more or less circular clump, which begins to show a slight differentiation into the component parts. There is no sign of interproglottidal glands. The posterior flap, of course, forms a dark line across the segment, but this is not homologous with the line of glands seen in *M. planissima*.

Segments 320^{mm} from the head show a very noticeable advance in the development of the sexual organs. The proglottids measure 3^{mm} wide by 0.8 long. No interploglottidal glands are visible. The posterior flap of each segment extends but slightly over the next following segment.

The genital pore has pierced the segment in the anterior third of the lateral edge. Cirrus-pouch and vagina are well developed; copulation has already taken place, as is shown by the fact that the receptaculum seminis is full of spermatozoa; the female glands form a large-rosette (diameter 0.56) just median of the lateral canals, the vitellogene and shell-glands being almost completely surrounded by the ovarium. The numerous testicles occupy the entire remaining part of the median field.

Segments 55^{cm} from the head measure 4^{mm} wide by 1.44^{mm} long. The uterus now fills the entire median field and a portion of the lateral fields, suppressing the other genital organs, with the exception of the receptaculum seminis and sexual canals. On these segments it could be proved that on one side the cirrus was ventral and the vagina was dorsal, while on the other side the cirrus was dorsal and vagina ventral. As the longitudinal canals could not be plainly seen, it is impossible to state which side is right and which left, but from analogy with the other forms it may be assumed that the side on which the vagina is ventral is right, the opposite side is left. The vagina is slightly posterior to the cirrus-pouch.

Segments 80^{cm} from the head measure 4^{mm} wide by 1.76^{mm} long. The uterus occupies almost the entire segment, and is filled with eggs. It could be proven that the uteri cross into the lateral fields only on one side (from analogy, dorsal side) of the large canal, and that the vagina passes on the same side, so that our analogy of ventral and dorsal, right and left sides, made above, is further supported. The dorsal longitudinal canals could not be distinguished. A transverse canal could be distinguished connecting the two ventral canals, thus making our analogy complete, and establishing the fact that we find the same topographical relations in this species which we find in *M. planissima*. The only point now left undetermined is whether the dorsal canal lies dorsal or ventral of the vagina and vas deferens.

Segments 1 meter from the head measure 5.5^{mm} wide by 3^{mm} long. The proglottids represent sacs filled with the uteri containing eggs; the pores are still plainly visible and a véry slight trace of the receptaculum seminis can still be seen.

Eggs.—The ova measure 60–68μ in diameter, the bulb of the pyriform body 16μ, the horns 8–10μ. Perroncito states that the eggs are cuboid, but that character is of course dependent upon reciprocal pressure, and hence of no diagnostic value. The measurements of the ova given by Perroncito and Neumann (in his Traité) are 48μ–52μ, by Moniez about 70μ.

SPECIFIC DIAGNOSIS.

The following may be given as a specific diagnosis of this parasite:

M. alba (Per., 1879) R. Bl., '91.—Whitish, 0.60ᵐ–2.50ᵐ long. Head, sub-quadrangular, 1.40–1.15ᵐᵐ broad; suckers slightly lobed, opening small and pointing anteriorly. Neck 1.5–5.3ᵐᵐ long by 0.6–0.9ᵐᵐ broad. Mature segments attain 8–14ᵐᵐ broad by 2–6.5 long by 1.5 thick. Topography of organs same as in *M. planissima;* testicles arranged in a quadrangle; genital pores in anterior half of the segment. Interproglottidal glands absent. Ova 60–88 μ; bulb of pyriform body 16–24 μ; horns 8–20 μ.

Type specimens with Prof. Perroncito, Turin, Italy; Bureau of Animal Industry, and private collection of C. W. Stiles, Washington, D. C.

C. W. S.

D. Moniezia. *(Undetermined specimens.)*

In the Bureau collection there are several very peculiar specimens of the genus Moniezia, which at present we are not able to determine specifically. They are all totally different from the forms described above, and it is believed they will be found to represent new species, at least new varieties, as soon as sufficient material can be obtained upon which to base an opinion.

No. 612, Bureau collection, contains a tapeworm from sheep. The interproglottidal glands are linear, but are very short. The head is enormous (Plate XVI, Figs. 5–5a), and totally unlike any head we have ever seen before. It is evidently closely related to *M. Benedeni* and *M. Neumanni.*

Nos. 725 and 732, Bureau collection, contain two heads and numerous loose segments of a worm of the *Expansa group.* The segments, in their microscopic anatomy, agree perfectly with *M. expansa,* but the general appearance of the head (Plate XVI, Figs. 4–4b) and segments is totally different from that species. The segments are very distinct, and appear almost like a string of large broad beads. However, a specimen of *T. saginata* has been found in the Army Medical Museum, the segments of which represent a perfect string of beads, so that we are inclined to doubt the specific value of the peculiar appearance of the segments of these specimens. The parasites came from Virginia sheep.

No. 607, Bureau collection (borrowed from Hassall's private collection), contains a strobila 54.5ᶜᵐ long by 2–2.75ᵐᵐ broad (Plate XVI, Figs. 1–2b). Interproglottidal glands could not be distinguished; specimen is poorly preserved. Genital pores are double; segments become longer than broad. This is the strobila referred to in the discussion of *T. Vogti.* (Cf. 84). Segments measure 2.5ᵐᵐ long by 2 broad, to 1.5 long by 2.75 broad. C. W. S.

GENERAL SUMMARY OF THE GENUS MONIEZIA.

From the above discussion it will be seen that the species *M. planissima, M. Benedeni, M. Neumanni, M. expansa, M. oblongiceps, M. trigonophora, M. alba,* and *M. denticulata* possess certain characters in common which compel us to follow Blanchard in uniting them in one genus. Some of these characters could not be traced in all the species owing to a paucity of material; nevertheless, the following will serve as a general summary of the anatomy of these forms.

The scolex is never provided with hooks. The segments are generally broader than long and longer than thick, although the end segments show a tendency to become longer and narrower, in some cases the length becoming equal to or (rarely) even greater than the breadth.

Genital organs.—The first appearance of the genital organs is in segments near the head, in the form of a round clump of tissue on each side of the segment, situated at the edge of the median field just inside the longitudinal canals. Gradually this genital anlage extends across into the lateral fields on the dorsal side of the lateral canals and nerves, assuming a characteristic pistol shape. It then becomes divided into an anterior (male) portion and a posterior (female) portion, which are at first solid, but in which canals gradually develop. The cuticle on the edge of the segment then invaginates at a certain point to meet the canals. At this point is situated the genital pore.

Male genitalia.—The anterior canal mentioned above differentiates, at its lateral end, into the cirrus-pouch; the remaining portion (vas deferens) becomes convoluted and runs towards the median line of the segment, passing on the dorsal side of the ovary. The testicles in many (all?) cases are at first arranged in two triangles, one on each side of the segment. This arrangement remains more or less constant throughout life in one species (*M. trigonophora*), while in (all?) the other species testicles appear in the remaining portion of the median field, thus forming a quadrangle. It is very rare that testicles are found in the lateral fields.

Female genitalia.—The lateral end of the posterior portion of the above-mentioned pistol-shaped genital anlage forms the vagina, the vulva being situated ventrally of the cirrus on the right side of the segment and dorsally of it on the left side. A receptaculum seminis is formed by the portion of the canal inside (median) of the longitudinal canals, while the extreme median lobe-like end differentiates into ovary, vitellogene gland, and shell-gland, together with the canals connecting the same as described on pages 20, 40. The female glands do not develop quite so early as the male glands, but they persist longer.

A uterus is present on each side of the segment. The two uteri form numerous folds, extending into the lateral fields (in all cases?, cf. *M. denticulata*), across the dorsal side of the longitudinal canals and nerves, and also toward the median line of the segment. Generally it is impossible to distinguish the uteri from each other in the median line of the older segments, but in *M. trigonophora* they more frequently remain distinct. As the eggs are developed and the uteri increase in size, the genital glands atrophy, the male organs disappearing before the female organs. Receptaculum seminis, vagina, cirrus-pouch, and the lateral portion of the vas deferens persist the longest. One segment has been found with a single median set of female organs and a single lateral pore.

The ova are spherical, but by reciprocal pressure they become cuboid. A pyriform apparatus surrounds the oncosphere.

Excretory organs of the segments.—On each side of the segments are found two longitudinal canals running from the head to the posterior extremity, crossing the genital canals on their ventral side. The large lateral canal represents the ventral canal; its cuticular lining is extremely thin; near the posterior border of each segment the ventral canals of the two sides are connected by a transverse canal possessing a thin cuticular lining. The other longitudinal canals lie medio-dorsal of the ventral canals, have a thick cuticular lining, and are not connected in the segments by any transverse canals.

Nerves in the segments.—On each side of the strobila there extends a longitudinal nerve from the scolex to the posterior extremity, running lateral of and parallel with the lateral longitudinal canals. It crosses the genital canals ventrally.

Interproglottidal glands.—A number of the species mentioned above possess glandular structures at the border between every two succeeding segments. The most simple and, as the writer considers, the primitive form of these structures, appears in species of the *Planissima* group, where the glandular cells are arranged in a line parallel to the posterior border of the segment. The longitudinal muscles cause a slight grouping of these gland-cells. In members of the *Expansa group* the gland-cells are concentrated around blind sacs, which open under the flap of the posterior edge. These glands generally appear in segments showing a genital anlage, although the exact distance from the scolex is subject to variation.

R. Blanchard included in this genus the following forms:

Tænia expansa R.; *Tænia alba* Per.; *T. Benedeni* M.; *T. denticulata* R.; *M. nullicollis* M.; *M. Neumanni* M.; all of which I also admit to the genus.

Blanchard further classes as *Moniezia*:

Tænia festiva R., 1819 (*M. festiva* R. Bl., 1891), from *Macropus giganteus*.

Dipylidium latissimum Riehm, 1881 (*M. Goezei* (Baird, 1853) R. Bl., 1891), from rabbits.

Dipylidium Leuckarti Riehm, 1881 (*M. Leuckarti* (Riehm, 1881) R. Bl., (1891), from rabbits.

Tænia marmotæ Frölich, 1802 (*M. marmotæ* (F.) R. Bl., 1891), from *Arctomys marmota*.

Tænia pectinata Goeze, 1782 (*Dipylidium pectinatum* Riehm, 1881; *M. pectinata* (G. 1782) R. Bl., 1891), from rabbits.

Moniez includes in the genus *Moniezia* the species:

Tænia ovilla Riv., 1878 (*T. Giardi* M., 1879; *M. ovilla* M. 1891), from sheep.

Thysanosoma actinioides Dies., 1834 (*Tænia fimbriata*, Dies.; *M. fimbriata*, Moniez, 1891) from sheep.

Blanchard gives as generic diagnosis the following: "Corps lancéolé en avant. Anneaux serrés, beaucoup plus larges que longs, avec deux pores sexuels opposés."

As the species *Tænia marmotæ, Dipylidium latissimum, D. Leuckarti,*

D. pectinatum, T. Giardi, and *Thysanosoma actinioides* have little in common with the species of *Moniezia* of cattle and sheep, it is proposed to exclude them from this genus.

Anticipating the results obtained below, it will simply be stated here that *Thysanosoma actinioides* and *Th. Giardi,* should be excluded from *Moniezia,* because in these two species we find in each segment a single uterus with ascon-spore like egg-sacs, instead of two uteri. Further, the genital canals pass between the longitudinal canals. For other particulars see the anatomy of these two forms.

Tænia marmotæ differs from *Moniezia* s. st., (1) in having a single uterus; (2) the dorsal canal lies lateral instead of median of the ventral canal; (3) the cirrus and vagina do not have the same topographical relations that are found in the genus *Moniezia.*

Dipytidium Leuckarti, according to Riehm,[*] has but one uterus in each segment, while (Riehm, Plate VI, Fig. 5) what is evidently the dorsal canal agrees with *T. marmotæ* in lying lateral of the ventral canal.

The uterus of *D. pectinatum,* according to Riehm, agrees with that of *D. Leuckarti.*

In *D. latissimum,* according to the same author, the uterus is not "ein einfacher, stellenweise erweiterter Schlauch, der in querer Richtung die Proglottis durchzieht, er erweist sich vielmehr in zwei oder drei solcher Schläuche zerspalten, welche sich vielfach mit einander vereinigen und so Inseln von Parenchym einschliessen." Riehm's figure of the uterus is very unsatisfactory, and resembles the vas deferens more than it does a uterus. Whether he has in this case drawn the uterus or not, the anatomy (to judge from his figure) does not agree with that of *Moniezia* s. st.

Hence it is proposed to restrict the genus *Moniezia,* and the following is suggested as a—

GENERIC DIAGNOSIS.

Moniezia R. Bl., 1891. Char. emend.—Head without hooks; segments generally broader than long and longer than thick, end segments showing a tendency to become longer and narrower. Two full sets of genital organs, with two uteri and two lateral pores in each segment. On the right side the vagina is ventral, cirrus dorsal; on the left side vagina dorsal, cirrus ventral. Dorsal canal lies dorso-median of ventral canal. Genital canals cross the longitudinal canals and nerves dorsally. Interproglottidal glands generally present. Calcareous bodies absent from parenchyma. Eggs with well developed pyriform body.

Besides the species mentioned above as belonging to this genus, there are several other forms which may or may not represent new species. This Bureau is not willing to commit itself in regard to the specific determination of specimens Nos. 607, 612, 725, and 732 of its collection, until further material is obtained. We incline, however, to the belief that at least two of these forms will eventually be proven to represent new species of the genus *Moniezia.*

For *M. nullicollis,* see p. 83.   O. W. S.

---

[*]Gottfried Riehm, Studien an Cestoden. Inaug.-Diss. Halle, 1881.

PART II.

THYSANOSOMA Dies., 1834.

(9) Thysanosoma actinioides Dies., 1834.

[Plate XI, Figs. 1-8.]

Synonymy.—*Thysanosoma actinioides* Diesing, 1834; *Tænia fimbriata* Diesing, 1850; *Moniezia fimbriata* (D.) Moniez, 1891; *Th. actinioides* Dies. (Reëstablished, Stiles. 1892.)

Hosts.—Sheep, (Curtice, Hassall, Stiles. *et al.*); *Cariacus rufus*, (Natterer); *C. simplicicornis*, (N.); *C. Nambi*, (N.); *C. paludosus* (*C. dichotomus*), (N.).

Geographical distribution.—South America: Brazil, "Registo do Rio Araguay, Cuyaba, Nas Frechas, und Villa Maria" (Natterer). North America: Colorado, Utah, Nebraska (Curtice); New Mexico (Codweiss, Curtice); California, Oregon, Utah (Collins, Curtice); Missouri (Stewart, Curtice); Washington, D. C. (sheep came from Colorado—Hassall and Stiles.)

LITERATURE.

(1) DIESING. *Tropisurus* und *Thysanosoma*, zwey neue Gattungen von Binenwürmern aus Brazil; Med. Jahrbuch d. oesterr. Staat. Neue Folge. vii, vide pp. 105–111, taf. III, 1834, *Thysanosoma actinioides*.

(2) NORDMANN. In Lamark, Anim. s. vert., 2 éd., III, 591.

(3) DIESING. Syst. Helm., I, p. 501, 1850, *Tænia fimbriata*.

(4) ———— Zwanzig Arten von Cephalocotyleen. pp. 32–33; Denkschr. d. M.-N. Classe der K. Akad. d. Wiss. Wien, 1856, *T. fimbriata*.

(5) KÜCHENMEISTER. On Tape and Cystic worms, with an introduction on the origin of the intestinal worms, by C. T. v. Siebold. Trans. by T. H. Huxley, 1857, p. 42.

(6) FAVILLE. Report Veterinary Dept. of the Colorado State Agricultural College, January, 1885, *T. expansa*, p. p.

(7) CURTICE, C. Tapeworm disease of sheep of the Western Plains; Annual Report of the Bureau of Animal Industry, U. S. Dept. Agric., 1887–1888 (appeared 1889); pp. 165–184, 2 plates, *T. fimbriata*.

(8) ———— Animal Parasites of Sheep, 1890; pp. 89–109, pl. XII (orig.), Pl. XIII (Diesing's figures), *T. fimbriata*.

(9) NEUMANN. Traité, 2 éd., 1892, p. 408, figs. 193–194 (after Railliet, unpublished), *T. fimbriata.*—Engl. Transl., p. 419.

(10) STILES. Bemerkungen über Parasiten—17: Ueber die topographische Anatomie des Gefässystems in der Familie *Tæniadæ;* Centralblatt f. Bakt. u. Parasitenkunde, bd. XIII, 1893, p. 457–465, figs. 7–8.

HISTORICAL REVIEW.

Natterer first discovered this parasite in *Cervus paludosus* in 1823. Only a few segments were found, and these were described by Diesing (1) under the name of *Thysanosoma actinioides*, for which he was inclined to erect a new order, *Craspedosomata*, to be placed between the

Cestodes and *Trematodes.* As Diesing misunderstood the nature of the animal, and mistook a single segment for the entire parasite, his description need not be reviewed.

In his Syst. Helm. (3) Diesing, having received complete specimens from Natterer, recognized that the animal was a tapeworm. He renamed the parasite *Tænia fimbriata* and gave as diagnosis:

Caput obtuse tetragonum magnum, acetabulis hemisphæricis angularibus anticis. Collum nullum. Corpus antrorsum lanceolatum, articulis cuneatis brevissimis, superiorum margine postico integerrimo, subsequentium crenato, ultimorum utriuque fimbriato. Aperturæ genitalium—. Long. 6′′′-6′′; lat. 1-3′′′.

In his third publication (4) he gives substantially the same diagnosis, adding "*articuli maturi soluti* 1′′′ *longi,* 2′′′ *lati, fimbriis lanceolatis.*"

Siebold's remarks (5) are simply of an historical nature.

The next article at my disposal is by Faville (6), who found tapeworms in sheep in Colorado. The worms were found in the gall-ducts and in the intestines. He includes all the worms under the name *Tænia expansa,* but there can be little doubt that the worms he found in the gall-ducts were *Th. actinioides,* while the worms he describes as 5 or 6 feet long must belong to some other species.

Curtice (7) had occasion to investigate a *Tænia fimbriata* epizoötic in Colorado in 1886 and 1887. He states that the adult specimens are 15–30^{cm} long by 8^{mm} wide. The broadest portion is about 2^{cm} from the posterior end. Head is 1–1.5^{mm} wide. Genital pores double. The segments attain maturity about the middle of the worm. The male organs develop first, and occupy the entire breadth of the young segments. The uteri develop last. Each uterus is composed of a series of bags, arranged side by side in a fringe, which extends along the anterior edge of the segment. The bags empty into a tube which is connected with the ovaries. The worms are found in the gall-ducts and duodenum. Curtice found one specimen in a lamb 2 months old, but he states that yearlings and 2-year olds are more infested than lambs. The worm requires six to ten months for its development. Curtice's experiments for direct development without change of host were negative. He further discusses the geographical distribution and treatment. He also refers to an article by Stewart. The reference is, however, incorrect and it has been impossible to trace the paper. Curtice reprints his observations in his report on the animal parasites of sheep (8).

Neumann (9) evidently takes his diagnosis from Curtice.

Stiles (10) remarks upon the anatomy and reëstablishes the genus *Thysanosoma.*

Besides the above articles this species is mentioned in several places, but no new facts are given in regard to its anatomy or life history.

Specific name.—According to the rules of nomenclature, the specific name given to a fragment of an animal must hold for the entire animal in case the fragment was named first. Hence nothing remains in this case but to accept the specific name *actinioides.* In separating this worm, together with *T. Giardi,* from the genera *Moniezia* and *Taenia*

s. st., it is necessary to accept the generic name *Thysanosoma* instead of creating a new name.

Considerable material is at hand which Curtice collected in 1886; it is, however, very poorly preserved, so that there are a number of points in regard to the structure of the parasite which it has been found necessary to omit.

ANATOMY.

The scolex.—The head (Plate XI, Figs 2, 2*b*) measures 1–1.5^{mm} wide; it is very prominent, much broader (almost square when viewed *en face*) than long, so that it looks almost like a crosspiece placed upon the anterior end of the strobila. The four suckers are very powerful, separated from each other by deep depressions (in other words, the suckers are prominently raised); each sucker measures about 0.48^{mm} in diameter; the opening is circular or slightly oval, and is directed half anteriorly at the four corners of the head. The scolex is quite sharply defined from the neck (Plate XI, Figs. 2, 2*a*), which measures about 0.9^{mm} broad and is very flat. The unsegmented portion is very short, or almost absent; 0.48^{mm} from the head the segmentation is quite distinct, while at about 1.5 from the head the posterior edge of the segments show short prolongations, which increase gradually in length as the segments grow older, forming the fimbriæ, so characteristic of this species.

Segments 20^{mm} from the head measure 2.4^{mm} wide by 0.24^{mm} long. No internal organs could be distinguished. No interproglottidal glands are present. The posterior edge of the segments is about 0.25^{mm} broader than the anterior edge; the flap extends over the anterior edge of the next following segment by 48 μ, and is divided into about 70 lobes which form a fringe work around the entire segment.

Segments 40^{mm} from the head measure 2.7^{mm} broad by 0.32^{mm} long. A genital anlage becomes visible on each side of the segment, just median of the longitudinal canals.

Segments 60^{mm} from the head (Plate XI, Fig. 3) measure 3.2^{mm} wide by 0.35^{mm} long; the fringe is somewhat longer in the median line (0.16) than on the edge (0.032).

In segments 80^{mm} from the head the genital anlage has extended into the lateral field towards the posterior half of the edge. The anlagen of the male and the female canals gradually separate from each other. The fringe on the posterior edge increases in length.

Segments 100^{mm} from the head (Plate XI, Fig 4) measure 4.8^{mm} wide by 0.35^{mm} long. The pores have pierced in the posterior half of the lateral margin. The male and female canals can be distinguished from each other. Just median of the longitudinal canals the vagina ends in a clump of tissue in which it was impossible to distinguish any differentiation into ovary, etc.; the vas deferens forms a number of convolutions; the posterior portion of the median field of every segment is occupied by the testicles, which generally appear slightly less numerous in the median line than elsewhere.

Segments 120ᵐᵐ from the head measure 5.1ᵐᵐ wide by 0.36ᵐᵐ long. The vagina and cirrus are very long, thin, and distinct; copulation occurs in these segments. Fig. 7 represents the penis in copulation; the genital pore is closed by contraction; the cirrus is inserted into the vagina.

Unfortunately, the material at hand is so poor that an exact description of many interesting points in connection with this species must be omitted for the present, and as *Th. actinioides* is rarely found in this part of the country, it is doubtful when a complete account of the anatomy can be given. The following points could, however, be established:

The genital canals pass between the longitudinal canals, as in *Th. Giardi*. The inverted cirrus and the vas deferens are quite convoluted, the cirrus-pouch long; the testicles are confined to the posterior half of the median field. The vagina generally passes on the dorsal side of the proximal portion of the cirrus-pouch, but the position of the vulva could be distinguished with less certainty; in many (all?) cases it is situated *anterior* to the cirrus in nearly the same frontal plane. As the vagina passes into the median field (Fig. 6) it runs ventrally across the convolutions of the vas deferens, which tend towards the dorsal surface. The ovaries, etc., lie just median of the longitudinal canals, but the topography of the various canals can not be given. Only a single uterus (Fig. 8) is present in each segment. This is like the uterus of *Th. Giardi*, and so far as observed it develops in the same manner.

The ventral canals are lateral and run parallel to the nerves; the smaller canals lie somewhat median of the lacunes, and are distinctly dorsal. At the posterior border of each segment there are two small (Fig. 5) but distinct transverse canals; the ventral transverse canal connects the lacunes (ventral longitudinal canals); the dorsal transverse canal connects the dorsal longitudinal canals. In connection with the double transverse canals it is worthy of note that the dorsal canal is not so rudimentary as in the other species described in this paper. From both dorsal and ventral transverse canals a system of extremely minute irregularly branching canals could be traced into the parenchyma of the segment.

Segments 140ᵐᵐ from the head measure 5.1ᵐᵐ wide by 0.4ᵐᵐ long. The fimbriæ are about as long as the segment. Segments 170ᵐᵐ from the head measure 5.1ᵐᵐ wide by 0.4ᵐᵐ long. Some of the fimbriæ are 0.4ᵐᵐ long.

SPECIFIC DIAGNOSIS.

The following is given as a specific diagnosis of this form:

Thysanosoma actinioides Dies., 1834.—Strobila 15–30ᶜᵐ long; head large, nearly square when viewed en face, 1–1.5ᵐᵐ broad, placed like a T on the neck. Suckers very large, prominent; openings large, elongated or oval, at the four corners, and directed forward or half forward. Neck is exceedingly flat dorso-ventrally, and quite broad. Segmentation begins almost immediately back of the head. The broadest segments measure 5–8ᵐᵐ wide by 0.4–0.6ᵐᵐ long, and are situated about 2ᶜᵐ from the posterior end. The posterior segments show a decided tendency to become longer and narrower. Mature segments attain a thickness of 2.2ᵐᵐ. The posterior flap of the segments is broken up into fimbriæ, which in the end segments attain the length of the segment itself. There are present in each segment two lateral genital pores, two ovaries, two vitellogene glands, but only one uterus. The latter is situated in the anterior portion of the median field, is composed of a small canal with numerous blind sacs, and surrounded by a thick fibrous tissue. The genital canals pass from the median field between the dorsal and ventral canals and dorsally of the nerve. The dorsal canals are somewhat smaller than the ventral canals and connected by transverse segmental canals. Ova measure ? in diameter; horns of pyriform body not developed. (No ova at my disposal).

Typical specimens in the Bureau of Animal Industry. Specimens will be sent to museums and specialists exchanging with this Bureau. -

C. W. S.

(10) Thysanosoma Giardi (Riv., 1878) Stiles, 1893.

[Plates VII Fig. 5, XII, XIII.]

Synonymy.—*Tænia ovilla* Rivolta, 1878; *Tænia Giardi* Moniez, 1879; *Tænia aculeata* Perroncito, 1882; *Moniezia ovilla* (Riv., Mon.) Moniez, 1891; *Moniezia ovilla* var. *macilenta* Moniez, 1891; *Thysanosoma Giardi* (Riv., Mon.) Stiles, 1893.

Hosts.—Sheep (Rivolta, Macerta, Moniez, Stiles); cattle (Perroncito, after Riv.).

Geographical distribution.—Italy (Rivolta, Perroncito, Macerta); Germany (R. Leuckart, at Leipsic); France (Moniez, at Lille; Stiles, at Paris). Not reported in America up to date.

LITERATURE.

(1) RIVOLTA. Di una nuova specie di Tænia nella pecora, *T. ovilla*; Giorn. Anat. Fisiol. e Patolog. degli Animali, Pisa, 1878, pp. 302–308, 3 figs.; (also, in Studi fatti nel Gabin. di Anat. patolog. di Pisa, 1879, p. 79, 3 figs.).

(2) MONIEZ. Sur le *Tænia Giardi* et sur quelques espèces du groups des inermes; Comptes rendus de l'Acad. d. Sc., 26 mai, 1879, pp. 1094–1096.

(3) PERRONCITO. I Parassiti, etc., Milano, 1882 (pp. 244–246, *T. ovilla;* pp. 246–247, *T. aculeata*).

(4) MONIEZ. Sur quelques types de Cestodes; Comp. rend. d. l'Acad. d. Sc., 6 Mars., 1882, p. 662 (*T. Giardi*).

(5) ZÜRN. Die tierischen Parasiten, Weimar, 1882 (pp. 198–199, *T. ovilla*).

(6) RAILLIET. Éléments de Zoologie médicale et agricole, Paris, 1886 (p. 261, *T. aculeata;* p. 264, *T. Giardi*).

(7) PERRONCITO. Trattato teorico-pratico sulle Malattie piu comuni degli Animali domestici, Torino, 1886 (pp. 236–237, *T. ovilla;* pp. 237–238, *T. aculeata*).

(8) NEUMANN. Traité des maladie parasitaires non-microbiennes, Paris, 1888 (p. 383, *T. aculeata, T. ovilla, T. Giardi*).

(9) MONIEZ. Les Parasites de l'Homme, Paris, 1889, (pp. 123–124, *T. Giardi*).

(10) NEUMANN. Observations sur les Ténias du Mouton; Comp. rend. d. l. Soc. d'Hist. Nat. d. Toulouse, 1891, 10 fev., 4 pages (*T. ovilla*).

(11) MONIEZ. Notes sur les Helminthes; Rev. Biolog. du Nord d. l., France, 1891, T. IV, Oct. (*Moniezia ovilla, M. ovilla* var *macilenta*).

(12) BLANCHARD. Notices helminthologiques, 2 sér. 7; Mém. d. l. Soc. Zool. d., France, 1891, pp. 443–446 (*T. ovilla*).

(13) NEUMANN. Traité, etc., 2 éd., 1892, pp. 408–409, 410; figs. 195–196, *T. ovilla.*— English translation by Fleming, London, 1892, pp. 419–421.

(14) ——— Sur la place du *Tænia ovilla* Riv. dans la Classification; Comp. rend.d. l. Soc. d'Hist. Nat. d. Toulouse, 2 mars, 1892, 3 pages.

(15) STILES. Notes sur les Parasites—13: Sur le *Tænia Giardi;* Comp. rend. d. l. Soc. d. Biol., Paris, 1892, pp. 664–665.

(16) —— Bemerkungen über Parasiten—17: Ueber die topographische Anatomie des Gefässytems in der Familie *Tæniadæ;* Centralblatt f. Bakt. u. Par., 1893, bd. XIII, p. 457–465, figs. 5–6.

HISTORICAL REVIEW.

Prof. Rivolta, of Italy, (1) seems to be the first person who noticed that this tapeworm differs from *M. expansa,* so frequently found in sheep, deer, cattle, etc. In 1874, while working on cestodes, he came across a peculiar form, which he decided did not belong to the species *T. expansa,* and which scarcely agreed with the species *T. denticulata.* He labeled the worm " *T. denticulata?*". Four years afterward he saw a parasite which Perroncito found in an ox and which he (R.) immediately recognized as identical with the form he had found in 1874 and labeled " *T. denticulata?*". Rivolta gave a very good description (1) of the worm, identifying most of the anatomical structures; he gave two figures of the parasite, which are easily recognizable, although unfortunately (evidently by a mistake of the editor rather than the author), the segments are turned upside down. For this new species Rivolta proposed the name *T. ovilla.*

Rivolta clearly recognized that the genital pores and ovaries were irregularly alternate; also that the testicles were placed laterally of the longitudinal canals. His description of the uterus is also quite exact. Rivolta's second paper (1879) is not at my disposal.

A year after Rivolta's publication Prof. R. Moniez (2), of Lille, France, described a tapeworm of sheep as a new species, to which he gave the name *Tænia Giardi.* He distinguished two ovaries in each segment, as is plainly seen from his statement that "Two of them (*i. e.,* currents of spermatozoa) are lost in the *neighboring ovary;*[*] the third extends across the segment and fecundates the *ovary*[*] of the *other side.*"[*]

Perroncito (3, 7) next mentioned *T. ovilla* in 1882 and 1886, adding nothing, however, to Rivolta's description. He also described as a new species (3, pp. 246–247) *Tænia aculeata,* which he considered as "very closely allied to, if not identical with, *T. ovilla.*"

Moniez (4) then mentioned *T. Giardi* again, but added scarcely anything new to his original description. In his book on the parasites of man (9) he states that the eggs of *T. Giardi* are arranged in bundles of 15–30 each, very similar to the egg bundles of *Tænia madagascariensis.*

Railliet (6) considered *T. ovilla* Riv. and *T. aculeata* Per. identical, and accepted the specific name *T. aculeata,* since *T. ovilla* had previously been used by Gmelin.

Neumann (10) then obtained type specimens of *T. ovilla, T. Giardi,* and *T. aculeata,* and decided that all three parasites belonged to one species,

---

[*] The italics do not exist in the original.

for which he accepted the name *T. ovilla*. A point which it is important to note is that Prof. Neumann states very decidedly that only one genital pore is present in each segment.

Since Neumann's paper appeared Moniez (11) has again written on this parasite. In this last paper he adopts the synonymy given by Neumann, and again asserts in most positive terms that double genital pores are present, and on this account he places the worm in Blanchard's new genus *Moniezia* (*M. ovilla*). ("Je me suis récemment encore assuré de son existence de la façon la plus certaine.") In the same paper Moniez describes as a new variety of sheep-tapeworm under the name *Moniezia ovilla* var. *macilenta*, a form in which ,the egg shells are small and the strobila is very thin.

R. Blanchard (12) has also recently published a paper in which he speaks of *T. ovilla*, and in which he states very positively that double genital pores were found in a specimen of *T. aculeata* fowarded by Perroncito, and of *T. Giardi* forwarded by Moniez (the specimen coming through the hands of Neumann).

At a meeting of the Washington Biological Society (April 16, 1892), the writer presented a paper on "*Tænia ovilla* and its relation to Blanchard's classification." In that paper, of which only an abstract (15) appeared in print, but which is substantially the same as the present paper, an attempt was made to harmonize the contradictory statements of the authors above reviewed. Since that date Prof. Neumann has sent me a reprint of a paper (14) presented him before the "Société d'Histoire Naturelle de Toulouse," and also a personal letter in which he reiterates the statements of his former papers.

Prof. Leuckart has also stated (personal correspondence) that a *T. ovilla*, with alternate genital pores, has recently been found in Leipsic, and Prof. Moniez has written (personal correspondence) in regard to the form in which he found the double genital pores.

When two experts like Blanchard and Moniez state that they have found double genital pores in *Tænia Giardi*, their statements must be accepted. Likewise, when Rivolta and Neumann state that they found alternate genital pores in *Tænia ovilla*, their statements must also be accepted, and we must try to harmonize the two opposing views. Two alternatives are open to us, *i. e.*, we can assume that some of the segments of *T. Giardi* have only one pore while the others have two pores, or that the forms examined represent two species. As will be seen by the text and figures given below, the former alternative is the correct one.[*]

---

[*] Since writing the above review, both Neumann and I have examined one of Moniez's slides which appeared to present two pores in one segment. Neumann, Moniez, and I now agree that in this particular slide the section is diagonal and intersects two segments. Thus this preparation does not prove that Moniez saw two pores. However, it seems to me highly probable that he saw double pores in some other specimen.—C. W. S.

Specific name.—The specific name *ovilla* must of course be dropped, for Railliet has already pointed out that *ovilla* was used by Gmelin[*] to designate another worm. Moniez's specific name *Giardi* must be accepted, since Neumann has demonstrated that this parasite is identical with *T. ovilla* Riv., and since the name *Giardi* antedates *aculeata* Per. The parasite certainly can not be united in one genus with such forms as *T. saginata*, nor can it be placed in the genus *Moniezia*. As shown below, it bears very close relations to *Thysanosoma actinioides*, and accordingly these two worms are placed in one genus.

ANATOMY.

The material studied was collected at one of the Paris abattoirs in the spring of 1891. The results harmonize perfectly with the statements of Rivolta and Neumann, in that as a rule the testicles of *Th. Giardi* are lateral and the genital pores are irregularly alternate. This rule, however, has two very important exceptions, *i. e.*, first, segments are not infrequently found in which double genital pores are present (hence the statements of Moniez and Blanchard); secondly, in some segments the testicles are not confined to the lateral spaces, but are also found in the median field, *i. e.*, in the space between the lateral canals.

Prof. Rivolta gives the total length of *T. ovilla* as 1.50^{m}; Perroncito's specimens of *T. aculeata* measured 2^{m}. I have examined a large number of segments, but no complete worm; the longest strobila was 1.20^{m}. The form of the segments varies greatly in different specimens, and according to the state of contraction there is considerable difference in the form of the segments of the same strobila, even when the segments are in the same stage of development. Plate XII, Fig. 1, represents a strobila about 1.20^{m} in length. It will be noticed that most of the segments are broader than long, while the end segments are slightly longer than broad. In the third and fourth portions it will be noticed that the lateral margins are irregular, owing to a slight lengthening of one side and the projection of the genital pore. This is an appearance which is frequent, though not constant. It is confined to the segments which are not full of eggs. Plate XII, Fig. 1a, represents segments in a stage corresponding to the last portion of Plate XII, Fig. 1. All the segments are broader than long, but the last segments show a tendency to grow narrower and longer, the same as in Plate XII, Fig. 1. The segments never attain any very great thickness; the anterior proglottids are quite thin. Plate XII, Fig. 7, gives a sagittal section of a mature segment. It measured 6^{mm} wide by 1.3^{mm} long by 0.19^{mm} thick. It will be noticed that the posterior edge of one segment overlaps the anterior edge of the one next following by about $96\ \mu$.

---

[*] Linné's Systema naturae, p. 3061. "*Tænia ovilla*, 20. *T. ovis arietis* Goeze, Eingew, p. 257."

Among the French tapeworms examined not a single head was noticed which could be positively identified as belonging to *Th. Giardi*. Two heads were found in a bottle containing segments of *Th. Giardi* and *M. expansa*, and it is probable that one of these heads belongs to the species now under consideration.

Perroncito (3, p. 246) describes the.scolex of *T. aculeata* as follows:

Testa depressa, quadrangolare, col lato maggiore di millimetri 0.560–0.589, col lato minore di millim. 0.250–0.300, lunga millim. 0.500; ventose rivolte in fuori, divise da una depressione, del diametro longitudinale di millim. 0.256–0.264, trasversale di millim. 0.200–0.208. Collo breve rappresentato da un restringimento o collaretto del diametro di millim. 0.536.

Neumann (13, p. 408) describes the head as "tétragone ayant un peu plus d'un millimetre de largeur, suivie d'un cou assez long, de largeur moitié moindre que celle de la tête."

One of the heads mentioned above agrees very nearly with the description, and is probably the head of *Th. Giardi*. It was connected with a strobila 38^{mm} in length.

The head (Plate XII, Fig. 2) is quadrate, stout, 0.506^{mm} broad, 0.33^{mm} long (from the anterior extremity to the constriction of the neck). The suckers are very strong, nearly round, with a diameter of 0.192^{mm}. The opening of the suckers in my specimen is elongated. The constriction back of the head measures 0.816^{mm}. Segmentation, at first extremely indistinct, begins at this point; 1.6^{mm} distal from this point the segmentation is very distinct. The first anlage of the genital organs is noticed about 2.25^{mm} back of the cervical constriction. At first only a small differentiation of tissue is seen in the median line of every segment; this arrangement (Plate XII, Fig. 3) continues for a distance of about 3^{mm}, when it is noticed that the anlage of every segment does not lie exactly in the median line, but slightly to the right or left of it. This diversion from the median line continues, so that 17.5^{mm} from the point where it begins the anlagen lie close to the longitudinal (lateral) canals. The oldest segment of this strobila measures 2.5^{mm} broad by 0.8^{mm} long. The genital anlage is very simple, and no traces of testicles can be seen.

Thus it is seen that this head and strobila belong to some cestode with alternate genital pores. If it is not a *T. Giardi*, it must be a new species (or *T. Vogti*???). The other head referred to agrees so closely with the head of *M. expansa*, that it is difficult to believe it can belong to *Th. Giardi*.

Plate XII, Fig. 4, represents seven young segments, each measuring 3.7^{mm} broad by 0.176^{mm} long. Six of these segments have a single anlage of genital organs, while one of the segments possesses anlagen of two sets of sexual organs, one of which is rather rudimentary. The genital pore has not yet pierced the cuticle of the margin. From about the middle of the lateral margin of each segment two curved and nearly parallel portions of tissue, which color darker than the surrounding parenchyma, extend across the longitudinal canal into the median field. The anterior string of tissue represents the anlage of the male genital canal, and is thicker in its marginal portion than in its median portion. The posterior (distal) string is the female anlage. The marginal portion is somewhat thinner than the corresponding portion of the male string; the portion which crosses the longitudinal canals is extremely thin and, upon arriving on the median side of the canals, ends in a small knob, which later differentiates into ovarium, vitellogene glands, etc. In the third segment of the figure, both male and female organs are seen on one side; and on the other side only the male anlage is present.

The segments of Fig. 5, Plate XII, measure (balsam preparation made under pressure) 3.5^{mm} broad by 1^{mm} long (median line); occasionally the edge of the segment upon which the genital pore is situated is slightly longer than the edge of the other side, so that the strobila has a zigzag appearance. The pore is in the middle or back of the middle of the margin.

In segments such as are shown in Fig. 5, Plate XII, we have the following arrangement of the genital organs. The genital pores are irregularly alternate, sometimes protruding prominently from the edge of the segment in the form of a papilla, at other times nearly on a line with the edge.

Male organs.—The lateral fields are almost entirely occupied by the male organs: the anterior portion (of the papilla-lateral field) by the vas deferens* and cirrus-pouch, the posterior portion by the testicles. The cirrus is occasionally seen protruding from the pore, at other times lying in the cirrus-pouch. The cirrus-pouch (0.32^{mm} long) is more or less bottle-shape, varying somewhat both in size and form in the different segments. Anterior to the cirrus-pouch the vas deferens makes a number of turns, then crosses the lateral canals into the middle field, extends diagonally across the ovary toward the posterior (distal) portion of the segment, and branches; one branch extends back to the testicles of the pore side, while the other extends toward the opposite side of the segment.

On the pore side about fifty testicles are found occupying the posterior portion of the field, and on the opposite side about ninety are found, but they occupy the anterior portion of the field as well as the posterior portion. The testicles are about 0.024^{mm} in diameter; they do not all lie in one plane, but by focusing can be proven to lie in both the dorsal and ventral portion of the segment.

Female genital organs.—The vulva is situated close to the cirrus; the vagina follows the general curvature of the cirrus-pouch to a point about equidistant from the anterior and the posterior margins of the segment, where it curves, crosses between the longitudinal canals diagonally, and enters the median field. Directly median of the longitudinal canals it dilates slightly into the receptaculum seminis, at the

---

* Neumann incorrectly figures testicles in this portion of the field.

end of which several organs come together, but in this stage it is scarcely possible to make out their exact relations. It can, however, be seen that a canal (ascending oviduct) extends toward the anterior edge of the segment crossing the ovary and forming a triangular dilatation. The base of this triangle extends parallel to the proximal edge of the segment in both directions towards the longitudinal canals. This, it will be seen later, is the anlage of the uterus. The branched ovary is seen at the point where the vas deferens crosses the oviduct. Posterior (distal) to the receptaculum seminis and ovary is a differentiation, which will be seen later to be the anlage of the vitellogene gland.

In segments (Plate XII, Fig. 6) measuring 3.4^{mm} broad by 0.8 long, the genital organs have increased in size, and the segments themselves are sharply defined from one another.

Plate VII, Fig. 5, shows a slightly diagrammatic drawing of the end portion of Fig. 6, Plate XII (microtome section). A small opening on the lateral margin of the segment leads into a sexual cloaca. The size of this sexual cloaca depends upon the position of the cirrus-pouch. If the cirrus-pouch is retracted into the body as in Fig. 6, Plate XII, the cloaca is quite deep, but if the cirrus protrudes the cloaca becomes inverted and forms what is called the sexual papilla (see Plate XII, Fig. 5). Thus the genital cloaca and genital papilla are not constant organs, but the one (in its fullest development) depends upon the absence of the other. When the papilla is present it is nearly filled up by the cirrus-pouch; when the cloaca is present the pouch begins at the deepest point of the former. The pouch is somewhat pear-shaped; the broader portion is situated towards the anterior margin of the segment; the narrower portion is slightly curved. The pouch of Fig. 6, Plate XII, measures 0.42^{mm} long. From the deepest point of the cloaca (highest point of the papilla) two canals extend towards the interior of the segment.

Male organs.—The anterior canal enters the tip of the pouch, in which it makes a number of irregular turns; the first portion is of nearly uniform size and provided with a lumen of 4μ, which is lined by a rather heavy cuticle. The wall of this portion is about 5μ thick. This is the cirrus or penis which is drawn into the cirrus-pouch. In one case the extruded penis was filiform, 80μ long. This canal gradually widens until the lumen measures about 25μ; the cuticle and wall become much thinner. The end of the broad canal suddenly dilates into a large vesicula seminalis, with a lumen of 60μ. From the vesicula seminalis the vas deferens, 64μ in diameter (lumen 8μ), extends into the anterior portion of the lateral field; the thick walls contain a tissue which is probably of glandular nature (prostata). The vas deferens decreases in diameter (32μ, lumen 8μ) as it extends towards the longitudinal canals; it passes on the dorsal side of the large canal (lacune of some authors) and on the ventral side of the small canal.

After passing the longitudinal canals the vas deferens runs across the ovarium, on the dorsal side of the latter. As stated above, it branches; this branching sometimes takes place before it reaches the ovary, but more frequently not until it has passed to the median side of the latter. The epithelial nature of the vas deferens is very evident, and at times an extremely fine cuticle (a basement membrane) could be distinguished. The testicles measure about 80μ in diameter.

Female organs.—The second canal, which extends from the end of the genital cloaca, represents the vagina. In the genus *Moniezia* we saw that this vaginal canal bears a constant relation to the cirrus-pouch, for on the right side of the strobila the vaginal opening and canal are constantly on the ventral side of the pouch, while on the left side they are dorsal. There seems to be no such constant topographical arrangement of the vaginal canal in the case of *Th. Giardi*, for although in Fig. 5, Plate VII, it is situated somewhat ventrally, it is plainly posterior (distal) in several segments of Fig. 5, Plate XII, and segments have also been found in which the canal was dorsal. The total thickness of the canal at the beginning is about 28μ; it increases gradually in thickness as it approaches the longitudinal canals, where it measures 40μ.

Zschokke states[*] that the vagina of *T. expansa* is composed of four layers: 1 (outside), a simple layer of circular muscular fibers; 2, a simple layer of longitudinal muscular fibers; 3, a homogeneous and brilliant tissue, and 4 (inside), a cellular membrane bearing cilia or hooks which project into the lumen of the vagina, and possibly serve in advancing the sperm cells into the receptaculum seminis.

In *Th. Giardi* I have been able to distinguish Zschokke's fourth layer very plainly, but have not been able to make out its cellular nature. The third layer is also very evident in my preparations. The first and second (outside) layers, which the eminent Swiss zoölogist describes in his excellent article on *T. expansa*, do not appear in my preparations of *Th. Giardi*. On the contrary, only one layer appears outside and that is distinctly cellular in its nature. These three layers end abruptly at the receptaculum seminis, which exhibits a distinct inner epithelial layer with a fine outer basement membrane.

Just median of the receptaculum seminis and connected with it is a small bulb in which the vitello-duct and oviduct come together; or, more properly speaking, the first portion of the oviduct extends from the calyx of the ovary toward the receptaculum seminis, on the median side of which it widens to receive the spermatozoa and vitelline matter. The ascending portion of the oviduct (not seen plainly on transverse sections, but very evident on frontal sections) then extends toward the anterior portion of the segment and empties into the uterus. The vitello-duct lies somewhat dorsally and distally of the descending oviduct, and empties into the bulb just referred to.

---

[*] Recherches sur la structure anatomique et histologique des Cestodes, p. 108.

It shows the same histology as the receptaculum seminis, and widens into a calyx around which are grouped the vitellogene cells. The gland in this stage measures about 0.16^{mm} in diameter. The cells of this organ color (in acid carmine) much darker than the cells of the surrounding parenchyma or the epithelial cells of the ducts. The descending oviduct agrees in histology with the vitello-duct, and widens also into a calyx around which the ovarial tubes are grouped.

The ascending oviduct is about 12μ thick, and consists of a simple epithelium with basement membrane. It passes dorsally of the ovarium, indenting the same, or even passing through its dorsal portion. The part of the oviduct between the ovary and the uterus is surrounded by a cellular tissue (shell-gland?) which colors about the same as the epithelium. The uterus in this stage is made up of a conglomeration of cells, but no lumen can be distinctly seen.

The *vascular system* drawn in Fig 5, Plate VII, is very instructive. Two longitudinal canals of different thickness and histological structure are seen. The external (ventral) canal presents a lumen of about 80μ, and near the posterior edge of the segment is connected with the corresponding canal of the opposite side by a transverse canal. The latter extends also into the lateral field of the segment. The lumen of the large canals (lacunes of some authors) is bordered by a fine membrane. Zschokke figures and describes a regular epithelium on the outside of this membrane, but in this species it has been impossible to recognize the epithelial nature of the neighboring cells, which seem to be exactly similar to the parenchymatic cells. These lacunes are supposed to be the ventral canals, a view which is supported in this species by the position of the genital organs—considering the female organs as ventral and the male organs as dorsal. The small longitudinal (dorsal) canals lie dorso-median of the lacunes. The diameter (24–36μ) varies at different points, the canal suddenly swelling into bulbs at various distances from each other.

The structure is entirely different from that of the ventral canals. The central portion, which is about 5μ broad, appears as a lumen when examined under a low power, but when examined under high power appears to be filled with a spongy mass in which no nuclei are present (coagulated contents?). This spongy centrum is bounded on the exterior by a 2μ cuticle which colors very dark. This is followed by a tissue (3μ) which colors much more lightly; the exact nature of this tissue is not clear. Next comes a distinct layer of circular fibers, which I am rather inclined to look upon as of muscular nature.

The longitudinal nerves run on the marginal side of the large canals and on the ventral side of the genital canals. In segments 4.25^{mm} broad by 1.04^{mm} long the organs are but little changed. The testicles measure 0.064–0.08^{mm} in diameter; the ovary is considerably larger than in the segments described before; the uterus is much thicker and has assumed a wavy appearance.

In Fig. 8, Plate XII, where the segments have reached a breadth of 5ᵐᵐ and a length of 0.72ᵐᵐ (somewhat shorter than the segments described last, a fact to which no importance should be attached, as this can easily be due to contraction) the " waves" of the uterus have advanced so far as to form a number of irregular parallel and connected folds. This peculiar development of the uterus is entirely different from the case presented by tapeworms belonging in the series with *Tænia solium*, *T. saginata*, etc. In these latter species the uterus develops lateral branches from a median stem, the median uterus not increasing in length disproportionately to the length of the segment. In the case of *Th. Giardi* the uterus lies parallel to the proximal edge of the segment; the two ends of the uterus are more or less stationary, and the uterus increases in length, totally disproportionately to the width of the segment, instead of sending off branches at right angles. The natural physical result of this growth is that the uterus lies in folds, which, in this case, assume a direction nearly parallel to the side of the segment. This mode of development of the uterus seems ample ground for separating, generically, worms of this kind from such forms as *T. solium*.

As the segments increase still further in size (Plate XIII, Fig. 2) (6ᵐᵐ broad by 1.04ᵐᵐ long) the folds of the uterus in growing larger are diverted from their position parallel to the sides and assume irregular positions; the testicles in the meantime become somewhat indistinct. In Fig. 3, Plate XIII (segments 6.5ᵐᵐ broad by 1.04ᵐᵐ long) the uterus occupies almost the entire median field, while the testicles are scarcely visible. The other sexual organs gradually atrophy.

Segments with more than one set of organs.—It has already been stated that Moniez and Blanchard have found segments with double pores, and that they believed that to be the normal condition. In Fig. 4, Plate XII, is pictured a segment with an extra male anlage. A number of segments have also been found in which an extra ovary and vitellogene gland were developed; also six segments in which double pores were present (Plate XIII, Fig. 1). One of these segments belonged to a strobila sent by Prof. Neumann. All the organs were double except the uterus, of which only one could be distinguished. With this the question as to the presence of single or double pores in *Th. Giardi* may be considered as definitely settled.

Eggs and egg-capsules.—Former authors have already stated that the eggs of *Th. Giardi* lie in capsules of fibrous tissue, each capsule containing 10–15 ova. Fig. 4, Plate XIII, shows the origin of these packages of eggs. At the top of the figure is seen a canal, which represents a portion of one of the slings of the uterus. Small blind sacs form on the side of this canal. The blind end of the sac enlarges, but still retains its communication with the uterus-stem through a narrow neck. The enlarged end is filled with eggs. The entire uterus with blind sacs is surrounded by a fibrous tissue, the layer being especially thick around the sacs.

Fig. 5, Plate XIII, represents these sacs in a further stage of development. Several eggs are seen at one end, but it is almost impossible to find the neck of the sacs. The eggs measure 20–23μ in diameter, and appear to have but one shell, as Moniez stated. I refrain from discussing the matter of the pyriform body, as no fresh material is at hand and preserved material is unsatisfactory in this case. Moniez (11) states:

Il faut considérer la coque chitineuse qui revêt immédiatement l'onchosphère comme homologue de l'appareil pyriforme, puisq' elle a la même origine. Le *Tænia ovilla* se trouvant donc rattaché aussi aux Anoplocephalines par le plus important des caractères de ce groupe, par l'existence de l'appareil pyriform autour de son embryon. * * *

This is, of course, a point of great importance.

SPECIFIC DIAGNOSIS.

From the above review of former work on *Th. Giardi*, and from my own anatomical description, we can accept the following as a diagnosis of this species:

Thysanosoma Giardi (Riv., '78) Stiles, 1892.—Length 1 to 2 meters. Head quadrate, measuring (after Neumann) slightly more than 1^{mm}; (after Perroncito) 0.56–0.589 (largest diameter), 0.25–0.3 (smallest diameter), 0.5 long; (after Stiles) 0.506 broad by 0.33 long (balsam preparation). Neck (after Neumann) rather long, half as wide as the head; (after Per.) short, diameter 0.536; (after Stiles) absent. The most of the segments are broader than long, thin; only the posterior are longer than broad, their breadth being less than that of the segment immediately preceding. Unripe segments present a slightly zigzag appearance, owing to the projection of the side on which the genital pore is situated. Largest segments 5–6.5^{mm} wide by 2–1^{mm} long. Genital pores in the middle or behind the middle of the segment and generally irregularly alternate, but occasionally segments are found with double pores. Testicles generally confined to the lateral fields, though scattered testes are occasionally found in the median field. Uterus transverse, giving rise to folds, which run more or less parallel to the lateral margin, and possessing ascon-spore-like egg sacs which are surrounded by an extremely thick layer of tissue. Genital canals pass between the dorsal and ventral longitudinal canals and dorsal of nerve. Eggs 20–23μ in diameter; a number are found together in small sacs of the uterus; horns of pyriform apparatus not developed.

Typical specimens with Bureau of Animal Industry and with Messrs. Neumann, Moniez, Rivolta, Leuckart, and Stiles.

THE SYSTEMATIC POSITION OF TÆNIA GIARDI.

It is evident from the above that this species can not remain in the genus *Moniezia*, since there is but one uterus present, and the topography of the genital canals, female glands and longitudinal canals is entirely different. It can not enter E. Blanchard's genus *Anoplocephala*, since the pores are irregularly alternate instead of unilateral. Neumann remarks that it approaches R. Blanchard's genus *Bertia*. This is true to a certain extent, but it is not deemed wise to place it in that genus until more is known about the position of the testicles, ovaries, etc., in *Bertia*. The occasional presence of double pores, on the other hand, shows that *T. Giardi* has some relations either with *Moniezia* or with *Thysanosoma*, and on account of its great resemblance to *Th. actinioides* in the uterus and in the topographical anatomy it may be provisionally placed in the genus *Thysanosoma.*

GENERAL CONSIDERATIONS IN REGARD TO TH. GIARDI AND TH. ACTINIOIDES.

From the foregoing discussion of these two species it will be seen that these forms can not possibly be united with the genus *Moniezia*, since they differ from the species of that genus already studied in a number of important characters, *i. e.*, the genital canals pass between the longitudinal canals instead of dorsal of both canals; the cirrus-pouch appears to be ventral of the vagina on both sides (*Th. actinioides*), instead of alternating dorsal right, ventral left; there is but a single uterus present in each segment, and this has an entirely different structure from that of the species of *Moniezia* described above.

In *Th. actinioides* the first anlagen of the genital organs appear on both sides of the median field of every segment, while in the anterior portion of what we have above supposed to be *Th. Giardi* (Plate XII, Fig. 3) the anlage first appears in the median line and then gradually diverges towards the lateral portion of the median field. Notwithstanding these differences the two worms are much more closely allied to each other than to the species of *Moniezia* above described, and should, for the present at least, be placed in a genus together. The following revised characters for the genus are proposed:

GENERIC DIAGNOSIS.

Thysanosoma Dies., 1834 (reëst. Stiles, 1893). Type species *Th. actinioides.*—Head unarmed; genital pores double or single; genital canals pass between the dorsal and ventral longitudinal canals and dorsal of nerve; only one uterus present in each segment; this uterus is transverse, waving, and forms blind egg-sacs which are surrounded by a thick layer of tissue; calcareous bodies absent from parenchyma; pyriform body of the eggs not well developed.

Th. actinioides and *Giardi* agree with each other in the relative position of the genital canals to the longitudinal canals, and in the form of the uterus. They differ from each other chiefly, however, in the position of the testicles and in the transverse canals, of which *Th. actinioides* has two, *Th. Giardi* one, to each segment.

Th. actinioides and *Th. Giardi* agree with *Tænia marmotæ* in having but one uterus. They differ from *T. marmotæ* chiefly in the relative position of the dorsal longitudinal canals, which in the latter species lie ventral of the genital canals and lateral of the ventral canals, and in the structure of the uterus.

Moniez states that *Tænia hyracis* possesses unilateral pores, and that the eggs are arranged in sacs. We know, however, too little about the exact nature of the sacs in that species as well as about the topographical anatomy of the other organs to determine in what genus the form belongs.   C. W. S.

(11) *Tænia marmotæ Fröhlich, 1802.

[Plate VII. Figs. 6, 7.]

Synonymy.—*T. marmotæ* Fröhlich, 1802; *Moniezia marmotæ* (F., 1802) R. Bl., 1891.
Host.—*Arctomys marmota.*

This species is introduced in this place for comparison, although it does not occur in cattle or sheep, because it is an important species to consider in connection with the genera *Moniezia* and *Thysanosoma.* Thanks to the kindness of Prof. R. Blanchard, I have been able to examine several specimens of *T. marmotæ.*

In the anatomy (Plate VII, Figs. 6, 7) of this species it is important to note that two pores are present in each segment. There are also two ovaries, two vitellogene glands, but only one uterus. This uterus is, however, totally different from that of *Thysanosoma,* for the branches or egg-sacs are not surrounded by the heavy layer of fibrinous tissue which is so characteristic of that genus. Furthermore, the eggs, according to Blanchard, are exactly like those of *Dipylidium latissimum* Riehm (*T. Goezei* Baird—*M. Goezei* (B) R. Bl.). In other words, the pyriform apparatus is well developed, which character does not agree with the genus *Thysanosoma.*

In *T. marmotæ* the topography of the longitudinal canals differs both from that of *Moniezia* and that of *Thysanosoma,* for the dorsal canal lies lateral of the ventral canal, and both of the canals as well as the nerve cross the genital canals ventrally, this latter character agreeing with *Moniezia,* but differing from *Thysanosoma.*

These characters seem to furnish sufficient grounds for separating the species *Tænia marmotæ* from the genus *Moniezia,* and for not placing it with the genus *Thysanosoma.* It certainly can not be united with such forms as *T. solium,* etc., on account of the transverse uterus, double pores, pyriform body, etc. It would also be impossible to place it in the genus *Dipylidium,* for it differs very greatly from *D. caninum* (*T. cucumerina* R.), which forms the type of that genus. The differences in the uterus, egg-balls, and rostellum are certainly sufficient grounds for not placing *T. marmotæ* in the same genus with *D. caninum.*

Should a new genus be established for this worm, as must undoubtedly be done, the generic diagnosis would read:

Head without hooks; segments broader than long. Each segment possesses two lateral genital pores; two ovaries and two vitellogene glands in the lateral portion of the median field; one transverse uterus with simple longitudinal branches. Dorsal canal between nerve and ventral canal; genital canals pass dorsally of longitudinal canals and nerve. Calcareous bodies absent from parenchyma. Eggs with well developed pyriform body.

Blanchard has already separated Riehm's species *Dipylidium Leuckarti, D. latissimum,* and *D. pectinatum* from the genus *Dipylidium,* and has placed them in the genus *Moniezia,* as *M. Leuckarti, M. Goezei,* and *M. pectinata.* The writer is in entire accord with Blanchard in the

_____

*Type. †—Typical specimens with Drs. R. Blanchard, of Paris, and C. W. Stiles, of Washington, D. C.

view that these three forms can not be generically united with *D. caninum*, but he can not admit them to the genus *Moniezia* (see p. 54), since—

(1) There is but one uterus present in the segments of *D. Leuckarti* and *D. pectinatum*, while the uterus of *D. latissimum*, although evidently different from that of *D. Leuckarti* and *D. pectinatum*, is not similar to that of *Moniezia*;

(2) What is evidently the dorsal canal in *D. Leuckarti* is figured by Riehm (Studien an Cestoden, Taf. VI, 5,) as lateral of the ventral canal. Riehm's species must be restudied before a positive decision can be made, but it seems very possible that these three forms can be placed in the same genus with *T. marmotæ* by slightly altering the generic diagnosis given above.   O. W. S.

PART III.

STILESIA Railliet, 1893

(12) Stilesia globipunctata (Riv., 1874) Railliet, 1893.

[Plate XIV.]

Synonymy.—*T. globipunctata* Riv., 1874; *T. ovipunctata* Riv., 1874; *Stilesia globipunctata* (Riv.) Rail., 1893.

Hosts.—Sheep (Rivolta, Giles); ? Cattle (See v. Linstow's Compendium.)

Geographical distribution.—Italy (Riv.); India (Giles).

LITERATURE.

(1) Rivolta. Sopra alcune specie di Tænia della Pecora. Pisa. 1874.

(2) Perroncito. I Parassiti, etc., 1882 (*T. globipunctata*, pp. 240, 241; *T. ovipunctata*, p. 241).

(3) ———. Trattato, etc., 1886, p. 235.

(4) Railliet. Éléments, etc., 1886, pp. 260, 261 (*T. globipunctata*, *T. ovipunctata*).

(5) Neumann. Traité, 1888, p. 384.

(6) ———. Observations, etc.; Compt. rend. d. l. Soc. d'Hist. nat. d. Toulouse, 1891.

(7) ———. Traité, 2nd ed., 1892, p. 410;—Engl. Transl., p. 420.

(8) Stiles. Bemerkungen über Parasiten—17: Ueber die topographische Anatomie des Gefässsystems in der Familie *Tæniadæ; C. f. B. u. P.*, 1893, XIII, p. 457.

(9) Railliet. Éléments, etc., 2nd ed., 1893 (MS).

HISTORICAL REVIEW.

Rivolta described *T. globipunctata* and *T. ovipunctata* in the same paper (1) in which he described *T. centripunctata.*   The article is not at hand.

Perroncito (2, 3) accepted both species; the following in regard to them is abstracted from his descriptions:

T. globipunctata (*globi-punteggiata*, Ital.). Head 1^{mm} in diameter; suckers directed forwards; neck absent. Length 45–60^{cm}; color white or greenish yellow, like the contents of the intestines. Ripe segments 2^{mm} wide by 0.15–0.17 long. Two spherical uteri are present; they appear as two lateral rows of whitish globules in the middle or posterior third of the strobila. Few eggs are present, oval, 28 μ by 24 μ; four hooks visible in the embryo.

T. ovipunctata (*ovipunteggiata*, Ital.). Head quadrangular, 0.50–0.65^{mm} broad; neck absent. Segments in sexual activity present to the naked eye two lateral lines composed of fine points, between which are seen larger bodies which are oval. This punctation is due in part to the vas deferens and in part to the uterus. The segments immediately following the head vary in breadth from 0.33–1^{mm}, with a length of 0.03; in mature segments the breadth is 1–2.5, length 0.08–0.12. Genital organs are on the sides of the segments; eggs oval, 20 μ by 16 μ.

73

The head of this worm buries itself in the intestinal mucosa and causes a degeneration of a number of the villi. It obstructs the function of the Lieberkühn glands to a certain extent, maintains an irritation in the cytogenic substance, giving rise to hyperplasia of the interglandular tissue, and forming nodules or elevations at those portions where the worm has fastened itself. In some places these elevations are confluent, in others isolated. Their size varies from that of a pea to a lentil, and they manifestly proceed from a circumscribed tumefaction of a fold of the mucosa.

Railliet (4) had evidently not examined the worms, but admits the two species. Neumann (5) adds that the testicles of these two species are in the lateral fields. He then (6) obtained a specimen which Rivolta had labeled "*ovi-globipunctata.*" Neumann states that the two species are identical, and that the median line remains transparent. In the second edition of his *Traité* (7) he holds to the same opinion, and unites the two worms under one diagnosis.

Von Linstow includes these worms in his list of cattle parasites, and following him the writer has done the same, although unable to trace any positive authority for so doing. Later (8) I stated that a new genus would have to be created for this and the following species.

Railliet (9) establishes a new genus *Stilesia* for *T. centripunctata* and *T. globipunctata.* (Personal correspondence; the second edition of Railliet's work is in press.)

Prof. Neumann very kindly sent a number of segments labeled *Tœnia globipunctata* which he had received from Prof. Rivolta, and Dr. G. M. Giles, of Sanawar, Punjab, India, has forwarded a number of specimens from India. Dr. Giles writes:

The worms are very social, dozens being found coiled up in a compressed mass in the small intestine. When fresh they are beautifully transparent and delicate. They would escape notice of anyone but a regular helminthologist, although masses half the size of one's fist may occasionally be found. In the fresh state they look so much like a mass of the pultaceous intestinal contents of this part of the intestine that the ordinary veterinarian would be pretty sure to overlook them.

I have found them in every sheep I have examined up here, but the fact that I have sometimes identified the worms as of one species, sometimes as another, makes me suspect that two or more species are present. *T. expansa* is also common in sheep out here, but not nearly so common as the delicate species. I have never found any tapeworms in the intestines of our oxen, i. e., *Bos indicus.*

It is upon these two sendings that the following description is based. Unfortunately the material was not preserved for fine microscopic work, so that the description given below is necessarily incomplete in many respects. Nevertheless, data enough are given to show that this worm is entirely different, not only specifically but also generically, from any of those described above.

ANATOMY.

General appearance.—The strobilæ in possession of this Bureau vary from 60–150ᵐᵐ in length, but no one strobila is complete. The widest segments are 2.5^{mm} broad, while the anterior and posterior segments are much narrower. In a fresh state they were transparent (Giles), and many of them have remained so in alcohol, while others have become more or less opaque. The head on the anterior end is quite prominent and appears like a knob. The anterior portion of the strobila is generally quite regular in outline, but the greater part of the worm is crenate and more or less twisted, thus making it quite difficult to prepare in mounts.

The head (Plate xiv, Figs. 2-2b) measures 0.768-0.9^{mm} in diameter. It is square when viewed *en face*. The suckers are generally quite prominent, and more or less lobed according to the state of contraction. They measure (balsam preparation) 0.336^{mm} in diameter, the muscular wall being 30μ thick. The opening is round or oval and directed in some cases anteriorly, in others more diagonally. The neck varies greatly according to contraction, in some specimens appearing to be absent; in others it is quite distinct. In one balsam preparation it measured 0.28^{mm} broad directly back of the head, and was 2.2^{mm} long. The longitudinal canals were very distinct, two lying dorsally and two ventrally, and measuring 6μ in diameter. The ventral canal was 16μ lateral of the dorsal canals. Segmentation becomes quite distinct at about 2.8^{mm} from the head. The distance between the dorsal and ventral canals gradually widens, and about 8.8^{mm} from the head irregularly alternate pistol-shaped genital anlagen appear, the handle of the pistol being situated in the space between the dorsal and ventral canals of the same side.

Segments 15^{mm} from the head (Plate xiv, Fig. 3) measure 0.56^{mm} broad by 56μ long. The longitudinal nerves were discovered in the extreme lateral portion of the segments. About 0.12^{mm} from the lateral edge ran the ventral canal, provided with an extremely thin cuticle and having a diameter of 8μ; 40μ nearer the median line was situated the dorsal canal, with a diameter of 6μ, and with a slightly heavier cuticle than that of the ventral canal. No transverse canals could be discovered connecting the dorsal canals, but near the posterior border of each segment there was a fine transverse canal running between the ventral canals. A short transverse canal also extended from each ventral canal towards the margin of the segment.

Male genital organs.—On each side of every segment are situated, between the ventral canal and the nerve, 4–7 testicles, measuring 8-9μ in diameter. The genital pore is not yet present. Irregularly alternate in the anterior portion of the segment there is a clump of tissue 68μ long by 12-14μ broad, which afterwards differentiates into the cirrus-pouch and the vagina.

Female genital organs.—From the posterior portion of this clump of tissue one can trace a thin string of nuclei which runs toward the longitudinal canals, crosses the ventral canal dorsally, and ends in a mass of tissue; the latter is situated between the dorsal and ventral canals, and represents the female glands; it is composed of an anterior and a posterior portion.

Segments 27^{mm} from the head (Plate XIV, Figs. 4, 5) measure 1.14^{mm} broad by 96μ long. The nerves have preserved their extreme lateral position and run ventrally of the genital organs. The ventral canals run about 0.18^{mm} from the lateral margin, and have a diameter of 16μ. The dorsal canals run 0.1^{mm} nearer the median line, and have a diameter of 7μ. The irregularly alternate genital pores have pierced in the anterior half of the margin. The genital cloaca is about 20μ deep, and from its proximal end two canals extend toward the median line. One of these canals (the vagina) lies dorsal, the other (the cirrus) lies ventral.

Male organs.—The cirrus-pouch is rather pyriform, 56μ long by 40μ broad. In its distal portion is situated the inverted cirrus, 50μ to 60μ long by 6μ broad. This colors very intensely, owing to the ciliated layer surrounding its lumen, and is thus quite sharply defined from the portion of the male canal which immediately follows. This latter makes a turn or two inside of the pouch, and then can be followed across the segment to the testicles of the other side. In its course it runs from the cirrus-pouch anteriorly of the testicles of the pore side, dorsally of the ventral canal and the female glands, ventrally of the dorsal canal, then running through the median field it lies anterior and dorsal of the transverse canal. It crosses the dorsal canal ventrally, the ventral canal dorsally, and is finally lost in the testicles. This vas deferens undoubtedly receives the spermatozoa from the testicles on both sides of the segment, although it was impossible to observe any fine branches running to the separate testicles of the pore-side. The testicles have increased considerably in size, and now measure 20–28μ in diameter.

Female organs.—The vulva lies on nearly the same frontal plane as the cirrus, or somewhat dorsally. The vagina then curves so as to lie dorsally of the cirrus-pouch. Ciliary projections could be seen for about 20–52μ from the vulva, then they became indistinct. The vagina then becomes very thin and crosses the ventral canal dorsally. At this portion it is generally somewhat convoluted. Just median of the ventral canal it increases somewhat in diameter, forming the receptaculum seminis. The median end of the latter then branches, forming two canals about 3μ in diameter. These two canals vary in their relative position. In general, however, one of them lies ventrally of the other, and after running a very short distance (about 15μ) toward the median line, it turns anteriorly and ends in the ovary. This canal is evidently the oviduct. The other canal extends just beyond the ovary, and is then lost to view. The ovary is a circular body 27–42μ in diameter.

In these segments one finds the anlagen of two more female organs, *i. e.*, the uteri. On the pore-side of the segment there is a clump of small darkly staining nuclei, situated ventrally of the receptaculum seminis and dorsally of the ovary. On the opposite side of the segment there is an anlage, which is exactly similar in appearance to the one just described.

Segments 50ᵐᵐ from the head (Plate XIV, Figs. 4, 5) measure 1.15ᵐᵐ broad by 96μ long. The ovary has increased in size, now measuring 60μ in diameter. The anlagen of the uteri have grown but little. For the next 8ᵐᵐ a complete and rapid change (Plate XIV, Fig. 6) takes place in the appearance of the female organs. It is noticed that the ovary now begins to grow less distinct, while the uteri come more plainly into view. We should naturally expect to find a canal connecting the descending oviduct and the uteri. No such canal, however, could be distinguished connecting the ovary with the uterus on the opposite side of the segment. In three different segments the writer thought he could distinguish a canal connecting the oviduct with the uterus on the same side, but was not positive that such was really the case; this point is left to be determined by some one who can procure and fix fresh material. It seems beyond question, however, that the ova descend through the oviduct and become fertilized; then, that a portion of them enter the uterus next to the ovary, while the remainder travel across the segment and enter the uterus of the other side.

In support of this view the following observations are given: The ovary becomes smaller while the uteri become larger; ova have been seen in the oviduct; at first no ova are seen in either uterus, then they appear in the uterus on the pore-side, and at the same time a broken line of ova can be seen extending across the segment toward the opposite uterus; the latter is next observed containing ova, while no ova can be distinguished in the median field. I am forced to admit that in no one segment has it been possible to find a continuous line of ova extending across the entire median field, but by combining several segments a diagrammatic line of ova may be constructed which extends from ovarium to uterus. An interesting point in connection with this wandering of the ova across the segment is that young ova, especially those found in the median field, have no definite form, a fact which points to their being capable of amœbic motion.

In segments measuring 1.344ᵐᵐ broad by 81μ long, still another change takes place in the form of the female organs. The uteri of both sides elongate medio-anteriorly, the prolongation having a fibrinous appearance. The oviduct is still faintly visible, and the ovary nearly or quite devoid of eggs can be distinguished on the ventral side of the uterus of the pore-side. At the same time another organ, the function of which I am unable to explain, appears anterior to the uteri and running transversely. This organ lies ventrally of the vas deferens—in those cases where the vas deferens extends so near the anterior margin—and stains quite dark in carmine or hæmatoxylin. The exact histology of this peculiar organ could not be definitely determined. In older segments the portion found in the median field looked very much like a canal.

In segments 1.256ᵐᵐ broad by 0.144 long each uterus is composed of two distinct portions, a posterior-lateral bulb, 96μ in diameter, containing the eggs which have now segmented and are surrounded by a shell, and an anterior median fibrinous prolongation 0.144ᵐᵐ long by 72μ broad at the base. The anterior extremity of this prolongation fits into (or connects with ??) the curve formed by the organ directly in front of it. This latter has increased considerably in size, and now extends partially around the uterus, crosses the dorsal canal ventrally, and tapers off into a fine point which runs through the median field to meet the corresponding organ of the opposite side. In one or two segments this crossed the dorsal canal dorsally instead of ventrally. The vas deferens which crosses the median field comes very plainly into view.

In segments 2.4ᵐᵐ wide by 0.114ᵐᵐ long the median prolongation of the uterus has partially constricted from the rest, so that we now find three more or less distinct portions of the uterus, i. e., the lateral bulb containing eggs and two spherical fibrinous portions. The middle portion lies nearer the ventral surface of the worm than either the lateral or the median portion. In different segments (Plate XIV, Figs. 7-9) these three divisions of the uterus present quite different relations to each other, but the exact order of the changes can not be given, as these variations were seen on loose segments and not on a complete worm. In the segments in which these changes are taking place the vagina is greatly enlarged.

As the segments grow older they become narrower and longer; the uteri assume a more antero-posterior position; the median anterior portion of the same becomes less distinct, and ova appear in the middle portion. The transverse organ in front of the uterus has united with the corresponding organ of the other side; the testicles become indistinct. Some of the oldest segments (Plate XIV, Fig. 9) of Dr. Giles' material measured 0.624ᵐᵐ broad by 0.28ᵐᵐ long.

In many segments one sees at the posterior edge, especially in the lateral fields, a row of small (4–8 μ) round or oval bodies which stain very dark in hæmatoxylin or carmine. These same bodies are occasionally met with in the median field, but their presence is extremely irregular. In general appearance they resemble, to a certain extent, the calcareous bodies found in *T. crassicollis, T. solium*, etc. In hæmatoxylin their margin and center stain very dark, while a lightly staining zone separates the two more darkly staining portions.

Eggs.—The eggs are round or oval, and measure 14–16μ x 10–14μ. Only one egg-shell could be discovered with certainty, although it occasionally appeared as if two were present; at two opposite poles this egg-shell bears a spinous projection 4–6μ long. Comparatively few ova (10–30) are found in each uterus.

SPECIFIC DIAGNOSIS.

The following is given as a provisional diagnosis of this species:

Stilesia globipunctata (Riv., 1874) Rail., 1893.—Strobila transparent, whitish or greyish yellow, 45–60ᶜᵐ long, very thin and not over 2.5ᵐᵐ broad. Head quadrate 0.5–1ᵐᵐ broad, suckers more or less distinctly lobed, opening round or oval and directed anteriorly or obliquely. Neck present, 2ᵐᵐ or more long (neck absent according to other observers). Developing segments 0.33–2.5ᵐᵐ broad by 0.05–0.17ᵐᵐ long; end segments narrower and longer (0.6 by 0.28). Genital pores irregularly alternate in anterior half of lateral margin. Cirrus-pouch slightly ventral of vagina, dorsal of nerve; vas deferens crosses the ventral canal dorsally, the dorsal canal ventrally; testicles in both lateral fields. Vulva on about the same frontal plane as cirrus; vagina generally dorsal of cirrus-pouch, crosses the ventral canal dorsally; ovary between the longitudinal canals on pore side of segment; two uteri present in each segment, one on each side and situated dorsally of ovarium; each uterus may be more or less distinctly divided into 3 globular bodies; a transverse organ of undetermined function anterior of uterus. At posterior border of lateral field is found a line of round or oval, refringent, chromatophil particles, 4–8μ in diameter. Ova 14–16μ by 10–14μ. Single shell with a spinous prolongation 4–6μ long at two opposite poles.

Type with Prof. Rivolta; typical specimens in Bureau of Animal Industry and collection of Stiles, at Washington; in collection of Leidy, at Philadelphia; and with Prof. Neumann, Toulouse, France.

It may be well to call attention to the following points, which should be more definitely determined by some one who is able to obtain perfectly fresh material: (1) What is the nature of the transverse organ anterior to the uterus? (2) Is there a canal extending from the descending oviduct to the uteri, or do the ova gain access to the latter by amœbic motion through the tissue? (3) What is the exact order of the changes in the form of the uterus? (4) What is the function of the canal which joins with the oviduct at the ovarium; do the ova pass through this canal to the median field? c. w. s.

(13) Stilesia centripunctata (Riv., 1874) Railliet, 1893 MS.

[Plate XV, Figs. 1–6.]

Synonymy.—*Tænia centripunctata* Rivolta, 1874; *T. centripunteggiata* (R.) Perr., 1882 (Italian); *Stilesia centripunctata* (Riv.) Rail., 1893.

Hosts.—Sheep (Riv., Mattozzi); ? Cattle (after ? see v. Linstow's Compendium).

Geographical distribution.—Italy (Riv. Mattozzi); Algeria (Neumann's specimens).

LITERATURE.

(1) RIVOLTA. Sopra alcune specie di Tenie della Pecora. Pisa. 1874.
(2) PERRONCITO. I Parassiti, etc., 1884, p. 242.
(3) ——— ———. Trattato, etc., 1886, pp. 235, 236.
(4) RAILLIET. Éléments, etc., 1886, p. 261.
(5) NEUMANN. Traité, 1 ed., p. 383.
(6) ——— ———. Observations sur les Ténias du Mouton; Compt. rend. d. l. Soc. d'Hist. Nat., 1891. 18 Mars.
(7) ——— ———. Traité, 2 ed., 1892, p. 409, fig. 197. Eng. Transl., p. 419.
(8) RAILLIET. Éléments, etc., 2 ed. 1893.

HISTORICAL REVIEW.

This parasite was first described by Rivolta (1) in 1874. The original article is not accessible. Perroncito (2, 3) says in regard to it that the head (scolex) measures 2^{mm}; suckers are large; neck entirely absent; the strobila measures 2.75–2.84^m long, wider in the anterior half than in the posterior half; 10^{cm} from the head the proglottids are 2–4^{mm} wide; 50^{cm} from the head they measure 2–3^{mm} wide; 150^{cm} from the head they measure 1.5^{mm} wide; at the end of the strobila they are scarcely 1^{mm} wide. They are quite thick, the thickness increasing as the segments grow older.

The mature specimens (preserved) were yellowish white and almost round. In the center of each segment, beginning at about the middle of the strobila, a round or elliptical and rather prominent spot could be noticed by the naked eye. This prominence increases in the posterior segments, alternates on the two surfaces, and is formed by the female organs in the center of the proglottid. The testicles are situated laterally in each segment; a single genital pore is present in each segment, situated in the middle of the lateral margin. Eggs are spherical with a single shell, diameters 22–24μ by 21–22μ; six hooks in protoscolex; very few ova present in each proglottid. Rivolta found these worms in the small intestines of a sheep. The mucosa was in part reddish, owing to hyperæmia of the villi, and in part pale and grayish on account of the pigmental degeneration of the villi. Dr. Mattozzi noticed only a catarrhal affection in the intestines of sheep infested with this parasite.

Railliet (4) takes his diagnosis from Rivolta or Perroncito, and has evidently not seen the parasite in question.

Neumann (5) does not state that he had examined the worms, but adds to Raillet's diagnosis that the length of the segment is $\frac{1}{10}$–$\frac{1}{4}$^{mm}, increasing with the age of the segments; the testicles are in the lateral portion of the median field; pores irregularly alternate.

Neumann (6) obtained several specimens from Algeria and adds to the foregoing that there are from 1,800 to 2,200 segments in each strobila. In the second edition of his *Traité* (7) he gives a figure of the head.

v. Linstow in his compendium places this species among the parasites of cattle. Following v. Linstow I also placed it in the Check-list of Animal Parasites of Cattle (Note 8), recently published, although I have been unable to ascertain when, where, or by whom it has been found in cattle.

ANATOMY.

Through the extreme kindness of Prof. Neumann I am in possession of a specimen which he obtained from Algeria, and although, as this French savant wrote, the strobila is not well preserved, the anatomy of this species will be described below as fully as the material allows. The scolex of the specimen is too contracted to warrant any statements (cf. Plate xv, Figs. 2–2b).

In a portion of the strobila measuring 1.04ᵐᵐ broad (Fig. 3) a darkly coloring tissue occupied the median line; on each side of this was a clear space; lateral from these spaces testicles were visible; then a clear space—a longitudinal canal; lateral of this there is another longitudinal line (nerve?).

In a portion of the strobila measuring 1.12ᵐᵐ wide (Plate xv, Fig. 4), instead of having a dark tissue on the median line, this portion of the field was clear for a space 0.16 broad. At the border of this field were found clumps of darkly stained tissue, irregularly alternate. These represent the handle of the pistol-shape genital anlagen; the muzzle of the pistol can be traced almost to the lateral edge of the strobila; further, it could be plainly seen that the testicles were situated not only on the median side of the longitudinal canals, but also dorsally (? or ventrally) and laterally of them.

The testicles measure 64μ in diameter (preparation was subject to pressure while coloring). In the strobila, measuring 1.24ᵐᵐ broad (Plate xv, Fig. 5), it is noticed that a line of deeply stained tissue runs from the handle of the pistol a short distance towards the opposite side of the segment; the lateral portion of the analage of the canals (muzzle of the pistol) is quite thick and partially differentiated into two parts (male and female). Where the strobila was 1.6ᵐᵐ wide (Plate xv, Fig. 6) the borders between the segments could be distinguished; the segments varied from 32μ long on the side without the pore to 64μ long on the side with the pore. The median portion (0.24ᵐᵐ broad) of the median field is now entirely occupied by the dark tissue referred to above as running from the handle of the pistol towards the other side of the strobila (probably uterus).

With the material at hand I do not feel justified in making any further statements.

SPECIFIC DIAGNOSIS.

The following is given as the diagnosis of this species:

Stilesia centripunctata (Riv., 1874) Rail., 1893. (*Provisional*).—Head large, 1.5–2ᵐᵐ broad, suckers large, directed anteriorly (after Neumann), situated at the four corners, directed diagonally forward (after Stiles). Strobila attains nearly 3ᵘ in length. Segments 10ᶜᵘ from the head measure 2–3ᵐᵐ wide, and from this point they grow narrower; posterior segments measure scarcely 1ᵐᵐ; always much broader than long, and longer on the pore side than on the opposite side. Genital pores irregularly alternate. Transverse uterus in the median portion of the median field. Testicles extend from each end of the uterus to the lateral (?) nerve. Eggs spherical with a single shell without pyriform apparatus; diameter 20–24 μ by 21–22 μ.

Type with Prof. Rivolta. Typical specimens with L. G. Neumann, Toulouse, Bureau of Animal Industry, and C. W. Stiles, Washington, D. C.

7114—No. 4——6

GENERAL REMARKS IN REGARD TO STILESIA GLOBIPUNCTATA AND STILESIA CENTRIPUNCTATA.

From the above description of these two species, although the account of *S. centripunctata* is very incomplete, it is evident that these forms can not be generically united with *Moniezia* or *Thysanosoma*. They differ from both of these genera in the form of the uterus and the general arrangement of the genitalia. *S. globipunctata* agrees with *Thysanosoma* in the relation of the genital canals to the nerves and longitudinal canals. This relation could not be definitely determined for *S. centripunctata*.

A new genus must certainly be established for *T. globipunctata*, and although *T. centripunctata* agrees with this form in comparatively few characters—narrow strobila, short segments, irregularly alternate pores, absence of testicles from median portion of median field, size of egg and single egg-shell—it will probably be better to place the two forms in one genus until *T. centripunctata* can be more thoroughly studied. Railliet has also recognized the necessity of establishing a new genus for these forms, and has proposed the generic name *Stilesia* (*S. globipunctata* and *S. centripunctata*). The following will suffice for a provisional diagnosis until *T. centripunctata* can be thoroughly studied:

GENERIC DIAGNOSIS.

Stilesia Railliet, 1893.—Type species *S. globipunctata* (Riv.) Railliet, 1893. Head with four suckers but no hooks. Strobila thin and narrow. Genital pores irregularly alternate. Segments broader than long. Two distinct sets of testicles present in each segment, one on each side, but no testicles in the median line. Eggs very small (not over 24 μ) and with but one shell.

The following points, which may prove to be of generic value, have been established only for *S. globipunctata:* Genital canals pass dorsally of nerve and ventral canal, but ventrally of dorsal canal. Egg-shell with two conical spinous projections at opposite poles.

Habitat.—Intestine of sheep. Development unknown.   c. w. s.

PART IV.

SPECIES INQUIRENDAE.

(14) Moniezia nullicollis Moniez, 1891.

[Plate XVI, 3-3b.]

Host.—Sheep (Moniez).

Geographical distribution.—Thus far *M. nullicollis* has been reported only at Lille, France, by Prof. R. Moniez.

LITERATURE.

MONIEZ. Notes sur les Helminthes VI, 3; Revue Biologique du Nord de la France, T. IV, 1891, 1 page.

HISTORICAL REVIEW.

Prof. Moniez found two specimens of tapeworms at the Lille abattoir in May, 1879. In 1891 he described them as a new species under the name given above, giving, however, only a very short description. He states that the worms were 40^{cm} long. The last segments were perfectly mature and measured, in alcohol, 8^{mm} wide by 1^{mm} long. They were quite thin and all the segments very short. Head cuboid, rounded in front, hemispherical in back, 1.3 in diameter. Suckers 0.65 in diameter. Neck absent, so that the segments begin directly back of the suckers. Ova 55-65μ; embryo 21μ.

At my request Prof. Moniez forwarded part of his original material. I have only the head (Figs. 3–3b, Plate XVI), and hence can give no statements in regard to the anatomy of the segments.

DIAGNOSIS.

It is impossible to give a satisfactory diagnosis of this form, hence I can not as yet look upon it as a well-established species.

Type with Prof. Moniez. One scolex in private collection of C. W. Stiles, Washington, D. C.   C. W. S.

(15) * Tænia Vogti Moniez, 1879.

Synonymy.—*Tænia Vogti*, Moniez, 1879; *Anoplocephala Vogti* Moniez, 1891.
Host.—Sheep (Moniez).
Geographical distribution.—France (at Lille, Moniez).

---

* Type specimen is lost.

LITERATURE.

(1) MONIEZ. Note sur deux espèces nouvelles de Tænias inermes (*T. Vogti et T. Benedeni*); Bulletin scientifique du Départment du Nord, 1878, II, pp. 163–164.

(2) RAILLIET. Éléments, etc., 1886, p. 261.

(3) NEUMANN. Traité, etc., 1st ed., 1888, p. 383.

(4). ————Observations sur les Ténias du Mouton; Comp. rend. d. l. Soc. d'Hist Nat. d. Toulouse, 1891, 18 Mars.

(5) MONIEZ. Notes sur les Helminthes; Rev. Biol. du Nord, 1891.

(6) R. BLANCHARD. Notices helminthologiques, 2 sér.; Mém. d. l. Soc. Zoöl. d. France, 1891, p. 447, footnote.

(7) NEUMANN. Traité, 2 ed., p. 408;—Engl. Transl., p. 419.

HISTORICAL REVIEW.

Moniez found a strobila 1½ feet long in a sheep, and gave the following description (1) of it: Head unknown; color of strobila, white; the young segments remain very narrow for a considerable distance of the total length; the largest segments measure 2.5^{mm} broad by 5^{mm} long, and contain eggs with the pyriform body; they are very flat; the muscles are better developed than in the other species, and are not grouped. It is very rare, having been found but once.

Railliet (2) and Neumann (3) take their diagnoses from Moniez. In (4) Neumann adds nothing to the above in his text, but in the analytical table he gives *T. Vogti* as having but a single genital pore to each segment. In Moniez's paper, however, I do not find any authority for this statement.

Moniez (5) then places this species in his analytical table as *Anoplocephala Vogti*, under the line " Un seul pore genital par anneau." This, however, is too indefinite, as the genus *Anoplocephala* in Blanchard's classification has unilateral pores.

Blanchard (6) calls attention to this last point, and states that this species is more closely allied to the genus *Bertia*, since the pores are irregularly alternate. I have been unable to obtain any specimens of this species.

CONCLUSIONS.

From this review it is seen that the *T. Vogti* is not sufficiently well known to be accepted as a true species. There is considerable doubt as to the genital pores, and our experience with *Th. Giardi* has taught us that great difference of opinion may arise in such a case, and that the pores alone are no criterion in the generic determination. Furthermore, the fact that the segments are longer than broad is by itself, unassociated with other characters, of little value, for we find that several species sometimes show this character.

In Hassall's collection is found a worm from sheep, the ripe segments of which agree in breadth with Moniez's description of *T. Vogti*, but the genital pores are distinctly double. In external appearance (Plate

XVI, Figs 1-2b) the worm is totally different from any tapeworm we have ever seen. While somewhat inclined to believe that this represents a new species—in case it is not identical with *T. Vogti*—I refrain from describing it as such, since the material is so poorly preserved that the internal organs will not stain. I am unable on this account to give any description of the internal anatomy, except that the vas deferens is very much convoluted, and that the uteri are present which resemble those of *Moniezia*. The time is passed when an author is justified in describing a tapeworm as a new species unless he can give a recognizable diagnosis.   O. W. S.

(16) *Tænia crucigera Nitzsch and Giebel, 1866.

Host.—Capreolus caprea.
*Geographical distribution.—*Germany, found but once, by Nitzsch.

LITERATURE.

GIEBEL, C. Die im zoölogischen Museum der Universität Halle aufgestellten Eingeweidewürmer nebst Beobachtungen über dieselben; Zeitschrift für die gesammten Naturwissenschaften, 1866, XXVIII, p. 259.

Giebel describes a species of tapeworm collected by Nitzsch from *Capreolus caprea*. His diagnosis and description are as follows:

T. capite obtuso obverse pyramidali tetragona collo brevi; articulis anticis tenuissimis, insequentibus latequadratis, ultimis subquadratis margine laterali arcuatis posteriori vix incumbentibus; vulvis oppositis.

"Im März zwei Exemplare von je 3 und 1½ Fuss Länge, letztes noch ohne abgehende Proglottiden. Der Kopf ist umgekehrt kegelförmig oder pyramidal, hinten am breitesten und fast wie abgeschnitten, mit vier hochrandigen mehr nach vorn als zur Seite gerichteten Saugnäpfen, die von vorn gesehen länglich viereckig, gleichsam mit eingedrücktem Kreuz bezeichnet sind. Der Halstheil hat halbe Kopfbreite und etwa die sech bis siebenfache Kopflänge. Von ihm nimmt der Leib ganz allmählig an Breite zu bis zum Ende, wo er vier Linien Breite misst. Die ersten Glieder gleichen bloss sehr feinen Runzeln, werden dann allmählig länger, anfang ¹⁄₁₀, dann ⅓-½-¾ zuletz ¾ ihrer Breite lang, gestreckt viereckig, am hintern Rande nur sehr wenig breiter wie am vordern. Der Seitenrand der vordern Glieder in 3 bis 4 Zoll Entfernung vom Kopfe erscheint durch zwei Hervorragungen gekerbt, deren vordere die Genitalwarze, die hintere die hervorstehende Hinterecke ist. Erste bildet ein sehr breites kurzes aufsitzendes Knöpfchen mit deutlicher Oeffnung. Weiter nach hinten wird die Geschlechtspapille schwächer, schon in der Leibesmitte ist sie ganz schwach und in dem letzten Leibesdrittel fehlt sie ganz. Der Hinterrand jedes Gliedes liegt hier nur ganz kurz und unbedeutend auf dem Vorderrande des folgenden auf, bei der nächst verwandten T. expansa überragt jener sehr beträchlich. Nahe des Seitenrandes der Glieder macht sich der Seitenkanal als dunkler Streif sehr bemerklich, nur in den ersten Gliedern nicht erkennbar. Die letzten Glieder sind ganz mit grossen reifen Eiern gefüllt, welche käseförmig gestaltet also in den breiten Seiten kreisrund, vom Rande gesehen länglich sind. Die Unterschiede dieser Art von der sehr ähnlichen vorigen sind folgende: Die Saugnäpe stehen sehr nach vorn und sind röhriger; der Halstheil ist deutlicher und länger, die hinteren Glieder in Verhältniss ihrer Briete weit länger; der Hinterrand der Glieder wenig oder gar nicht überragend; die Geschlechtspapille knopfförmig, die Eier käseförmig, nicht konisch."

---

* Type specimen is lost.

From the above it is very probable that this species belongs to the genus *Moniezia*, although, without the internal anatomy, even this point is uncertain.

Dr. Brandes, in reply to a letter, has written that the type specimen has, unfortunately, been lost or destroyed. It would be too uncertain to determine a worm by the above description without the original material at hand, so that it will probably be better to ignore this specific name and diagnosis entirely. C. W. S.

(17) Tænia capreoli Viborg, 1795 and (18) Tænia capræ Rud., 1810.

These two species should evidently be entirely ignored, for the descriptions given are not sufficient to establish them as true species, and the types can not be found. The following remarks in regard to them are quoted from Rudolphi's Historia Naturalis (pp. 200–201). The forms have been referred to by several authors since Rudolphi's time, but no one seems inclined to accept the species or able to give any anatomical details in regard to them.

83. TÆNIA CAPREOLI.—Viborg ind. Mus. Vet. Hafn., p. 238. n. 87. b. Tænia capreoli. *Tænia ovina (expansa* n. 2.) cl. viro n. 87. a. dicitur, unde capreoli vermem huic affinem esse, eo magis colligimus, quo major inter expansam (ovis), denticulatum (bovis) et insequentem (capræ) affinitas observetur.

84. TÆNIA CAPRÆ.—*Hab.*—In *Capræ Hirci* intestino ileo specimina plurima, mortua, Augusto mense reperi.

Fragmenta tri- vel quinquepollicaria; quorum alia minora articulis linea angustioribus, brevissimis, angulis laterlibus acutiusculis, alia majora articulis duas treave lineas latis, vix lineam longis. Isti crassiusculi marginibus lateralibus crenulatis, foramine utrinque medio sive opposito indistincto, angulis rotundatis, margine postico plicatulo, substantia molli, lineis transversis undulatis.

Obs. 1.—Tænia hæc inter expansam n. 2. et denticulatam n. 3. quas conferas, media videtur, capite tamen non viso, de eadem judicare nequeo.

Obs. 2.—Ne cum Tænia capræa Abilgaardii, caprina Gmelini et Zederi confundas, hæc enim Tænia non est, et sub Polystomatis denticulati nomine supra descripta sistitur. C. W. S.

(See also addenda, p. 101.)

PART V.

LIFE HISTORY.

Mégnin published an article in which he set forth the theory that certain cestodes could develop without a change of host. Moniez and others combatted this idea, while Perroncito suggested that some insect might possibly play the role of intermediate host for forms like *Moniezia expansa.* McMurrich was inclined to look upon *Melophagus ovinus* as the intermediate host of *Moniezia expansa.* Leuckart has suggested that some snail might act as the intermediate host. Curtice then supported the direct infection theory again.

Hutchinson, in rather an undecided amateurish article, suggests that it is possible that caterpillars, slugs, earthworms, lice, etc., may be intermediate hosts for *T. expansa,* which he "believes" is the tapeworm found in sheep and ruminants.

For over a year past we have been conducting experiments in this line, but up to date only negative results can be reported. Experiments to infect sheep directly by feeding eggs to them have been totally negative. Experiments to infect *Melophagus ovinus,* numerous coprophagous insects, and earthworms, have also been negative. It is the intention to continue these experiments, especially those with earthworms, this coming spring upon a much larger scale, and we hope later to report some positive results. From experiments thus far, however, we feel confident that cattle, etc., can not become infested with tapeworms by swallowing the eggs, as Curtice still contends, and it is believed that some insect, worm, or snail will be found to contain the larval stage. C. W. S.

PART VI.

CONCLUSIONS.

In the foregoing studies the following conclusions have been reached:

(1) Descriptions of cestodes based upon external form alone, unassociated with internal anatomy, are of little value. (See also p. 84, 99.)

(2) In the case of those forms of which the original type specimens can be obtained, we may be able to recognize the species, but in most of the cases of which the types have been lost, the determination of the species is extremely uncertain. We shall hence be obliged to ignore a great number of forms which have been described as species.

(3) The present genus *Tænia*, as generally accepted by authors, contains forms which must be restudied and arranged in several sub-families and a number of genera.

(4) This revision must be based upon internal anatomy.

(5) In the adult cestodes of cattle and sheep three genera *Moniezia* R. Bl., *Thysanosoma* Dies., and *Stilesia* Rail., are recognized.

(6) For a summary of the genus *Moniezia*, see p. 51; for *Thysanosoma*, p. 70; for *Stilesia*, p. 82.

(7) The following table will aid in determining the various species of these genera:

```
    ⎧ Uterus double; genital pores double; eggs with well de-
    ⎪    veloped pyriform apparatus....................MONIEZIA R. Bl. 2
    ⎪ Uterus single, transverse, with ascon-spore like eggs-sacs;
    ⎪    genital pores double or irregularly alternate; horns
  1 ⎨    of the pyriform apparatus not developed...........THYSANOSOMA Dies. 10
    ⎪ Uterus single or double and without ascon-spore like
    ⎪    egg-sacs; genital pores irregularly alternate; eggs
    ⎪    with a single shell; strobila narrow; testicles absent
    ⎪    from median portion of median field.................STILESIA Rail. 11
    ⎩ Species inquirendæ ..........................................9, 12

    ⎧ MONIEZIA. For generic diagnosis, see p. 54.
  2 ⎨    Interproglottidal glands absent ...............................  3
    ⎩    Interproglottidal glands present...............................  4

  3 ⎧ Genital pore in anterior half of lateral margin...............M. alba, p. 51
    ⎩ Genital pore in posterior half of lateral margin.............M. denticulata, p. 46

  4 ⎧ Interproglottidal glands linear, not grouped around blind
    ⎨    sacs ..........................................................  5
    ⎩ Interproglottidal glands grouped around blind sacs...........  7
```

88

5 { Interproglottidal glands very prominent and quite long
  (broad in reference to the segment); head not very
  distinctly lobed; opening of suckers slightly elon-
  gated; segments attain 20-26ᵐᵐ in breadth.............. *M. planissima*, p. 22
  Interproglottidal glands short and not prominent............................ 6

6 { Head very distinctly lobed and sharply separated from
  neck; opening of sucker round; neck nearly as broad
  as head; segments attain 10-12ᵐᵐ in breadth............. *M. Benedeni*, p. 25
  Lobing of head less distinct; head not so sharply sepa-
  rated from neck; neck filiform; segments attain 8ᵐᵐ
  in breadth; thinner than foregoing species.............. *M. Neumanni*, p. 26

7 { Testicles generally in form of two triangles; space frequently left between the uteri in the median line;
  head pyriform, almost square when viewed *en face;*
  segments attain 6ᵐᵐ in breadth; end segments occa-
  sionally as long as broad.............................. *M. trigonophora*, p. 42
  Testicles generally in form of a quadrangle; no space be-
  tween the fully developed uteri; end segments never
  as long as broad 8

8 { Head oblong when viewed *en face;* not lobed; suckers
  not distinctly raised; entire strobila thin; segments
  attain 9 mm. in breadth; end segments narrower .... *M. oblongiceps*, p. 36
  Head more or less distinctly lobed; nearly square when
  viewed *en face;* suckers raised; openings decidedly
  elongated; strobila often quite thick; posterior portion frequently of an orange color; segments attain
  16 mm., perhaps more, in breadth; end segments show
  a tendency to break off 1—3 at a time .... *M. expansa*, p. 34

9 { Neck absent; head large, decidedly lobed; openings of
  suckers round; segments attain 8 mm. in breadth—sp.
  inq. .... *M. nullicollis*, p. 83.

10 { Thysanosoma.—For generic diagnosis, see p. 70
  Head very large (1.5 mm); square; lobed; testicles in
  median field; posterior flap of segments fimbriate;
  genital pores double.................................. *Th. actinioides*, p. 68
  Head small; testicles in lateral fields; posterior flap not
  fimbriate; genital pores irregularly alternate (rarely
  double)... *Th. Giardi*, p. 69

11 { Stilesia.—See p. 82.
  Median portion of median field occupied by a transverse
  uterus; "head 2 mm. in diameter"..................... *S. centripunctata*, p. 81.
  Median portion of median field transparent; two lateral
  uteri in each segment; "head less than 1 mm. in dia-
  meter"... *S. globipunctata*, p. 79.

12 { Segments 2 mm. broad by 5 mm. long, sp. inq............ *T. Vogti*, p. 83.
  End segments 8 mm. broad, sp. inq .... *T. crucigera*, p. 85.

(8) For *Tænia marmotæ* we shall be compelled to recognize a new
genus (p. 71).

(9) Riehm's species *Dipylidium Leuckarti, D. pectinatum* and *D. latissimum* can not remain in the genus *Dipylidium*, established by Leuckart
for *Tænia cucumerina.* It may be found upon further study that they
can be placed in the genus with *T. marmotæ.*

O. W. S.

PART VII.

COMPENDIUM OF SPECIES ARRANGED ACCORDING TO THEIR HOSTS.

I have personally examined those forms marked with an asterisk (*) under the various hosts, and can hence guarantee that these species occur in the animals cited. The other cases mentioned are compiled. Forms given as *M. expansa*, but not starred, should be reëxamined to establish their specific identity.

The nomenclature of the hosts here adopted is mainly that of Flower and Lydecker's Mammals, Living and Extinct, 1891. In the case of the *Cervidæ*, Sir Victor Brooke's Monograph (Proc. Zool. Soc., London, 1878 pp. 883–928) of this family has been followed:

Antilope dorcas Pallas. See GAZELLA DORCAS.
Antilope rupicapra Erxl. See RUPICAPRA TRAGUS.
BOS INDICUS. Zebu.
 Moniezia expansa (R.) R. Bl., reported as *Tænia expansa.*
BOS TAURUS L. Domestic cattle.
 * *Moniezia alba* (Per.) R. Bl.
 * *Moniezia Benedeni* (M.) R. Bl.
 * *Moniezia denticulata* (R.) R. Bl.
 * *Moniezia expansa* (R.) R. Bl.
 * *Moniezia planissima* S. & H.
 Thysanosoma Giardi (Riv). S. (See p. 59.) Found by Perroncito; determined by Rivolta.
 Stilesia centripunctata. (See p. 79.)
 Stilesia globipunctata. (See p. 73.)
CAPRA HIRCUS L. Goat.
 Moniezia expansa (R.) R. Bl., reported as *Tænia expansa.*
 Tænia capræ R. sp. dub.
Capra hispanica Græl. See CAPRA PYRENAICA.
CAPRA PYRENAICA Schinz. (*Capra hispanica.*) Spanish Ibex.
 Moniezia expansa (R.) R. Bl., reported as *Tænia expansa.*
CAPREOLUS CAPREA. (*Cervus capreolus.*) Roe or Roe deer.
 Moniezia expansa (R.) R. Bl., reported as *Tænia expansa.*
 Tænia crucigera Nitzsch, sp. dub.
CARIACUS (BLASTOCERUS) CAMPESTRIS. (*Cervus campestris.*) Pampas Deer.
 Moniezia expansa, reported as *Tænia expansa.*
CARIACUS (BLASTOCERUS) PALUDOSUS. (*Cervus paludosus.*)
 Thysanosoma actinioides Dies.
CARIACUS (COASSUS) NAMBI. (*Cervus nambi.*)
 Moniezia expansa (R.) R. Bl., reported as *Tænia expansa.*
 Thysanosoma actinioides Dies.

90

CARIACUS (COASSUS) RUFUS. (*Cervus rufus.*) Brocket.
    Moniezia expansa (R.) R. Bl., reported as *Tænia expansa.* (See Diesing.)
    Thysanosoma actinioides Dies.
CARIACUS (COASSUS) SIMPLICICORNIS. (*Cervus simplicicornis.*)
    Thysanosoma actinioides Dies.; found by Natterer.
CARIACUS (COASSUS) SP. ?
    * *Moniezia oblongiceps* S. & H.
Cervus campestris. See CARIACUS (BLASTOCERUS) CAMPESTRIS.
Cervus capreolus. See CAPREOLUS CAPREA.
Cervus dichotomus. See CARIACUS PALUDOSUS.
Cervus nambi. See CARIACUS NAMBI.
Cervus paludosus. See CARIACUS PALUDOSUS.
Cervus rufus Cuv. See CARIACUS (COASSUS) RUFUS.
Cervus simplicicornis. See CARIACUS SIMPLICICORNIS.
GAZELLA DORCAS. (*Antilope dorcas.*) Gazelle.
    * *Moniezia expansa* (R.) R. Bl., one of Rudolphi's original specimens.
OVIBOS MOSCHATUS (Zimm). Musk ox.
    Moniezia expansa (R.) R. Bl.
OVIS ARIES L. Domestic sheep.
    * *Moniezia alba* (P.) R. Bl.
    * *Moniezia Benedeni* (M.) R. Bl.
    * *Moniezia expansa* (R.) R. Bl.
    * *Moniezia Neumanni* Moniez.
    * *Moniezia nullicollis,* Moniez, sp. dub.
    * *Moniezia planissima* S. & H.
    * *Moniezia trigonophora* S. & H.
    * *Thysanosoma actinioides* Dies.
    * *Thysanosoma Giardi* (Riv.) S.
    * *Stilesia centripunctata* (Riv.) Rail.
    * *Stilesia globipunctata* (Riv.) Rail.
RUPICAPRA TRAGUS Gray. (*Antilope rupicapra.*) Chamois or Gemse.
    Moniezia expansa (R.) R. Bl., reported as *Tænia expansa.*

C. W. S.

PART VIII.

BIBLIOGRAPHY OF ADULT CESTODES OF CATTLE AND SHEEP.

We have not been able to find the articles the numbers of which are inclosed in brackets, thus [ ]:

1. D'ARBOVAL, L. H. J. Helminthes; Dictionnaire de Médécine de Chirurgie et d'Hygiéne Vétérinaires, par. A. Zundel. Paris, 1875, pp. 111-159.
   (*T. expansa*, p. 150; *T. denticulata*, p. 151.)

2. BAILLET. Histoire Naturelle des Helminthes des principaux mammifères domestiques. Paris, 1866. (Extr. d. Nouv. Dict. d. Méd., d. Chir. et d'Hyg. vét.)
   (*T. expansa*, p. 161; *T. denticulata*, p. 161; *T. capræ*, p. 162.)

[3] BATSCH, A. J. G. C. Naturgeschichte der Bandwurmgattung überhaupt und ihrer Arten insbesondere, nach den neuern Beobachtungen in einem systematischen Auszuge. Halle. 1786.

4. BLANCHARD, RAPHAEL. Sur les Helminthes des primates anthropoïdes; Mémoires de la Société Zool. de France, 1891.
   Vide p. 187, footnote; (*Moniezia* mentioned as a new genus).

5. ———. Notices Helminthologiques. (Deuxiéme série.) 7. Cestodes du groupe des Anoplocephalinæ, p. 443; Mémoires de la Société Zoologique de France, 1891.
   (pp. 444-446. Generic diagnosis of *Moniezia*: (1) *Moniezia alba*, (2) *M. Benedeni*, (3) *M. denticulata*, (4) *M. expanza*, (9) *M. Neumanni*, (10) *M. nullicollis*. T. ovilla, T. fimbriata.)

[6.] BLOCH, M. E. Abhandlung von der Erzeugung der Eingeweidewürmer und den Mitteln wider dieselben. Berlin, 1782. 4°, 10 taf.

6a. ———. Traité de la génération des vers des intestins et des Vermifuges. Strasbourg, 1788.
   (*Tænia vasis nutriciis distinctis*, pp. 16-17, v, 1-5.)

[7] CAMPER. Beschäft. Berlin Naturf. Freunde iv, 1779. 39 *T. denticulata*.

8. CARLISLE, ANTHONY. On the structure and occurrence of Tæniæ; Transactions of the Linnæan Society (Lond., 1793), ii, tab. xxv, f. 15, 16.
   [*T. ovina*=*T. expansa*, after Rudolphi].

[9] CHABERT. Traité des Maladies Vermineuse, p. 120, pro parte.
   [*Tænia rubané*=*T. denticulata*, after Dies.]

10. COBBOLD, T. S. Parasites; a treatise on the entozoa of man and animals. London, 1879. *Vide* p. 332.
    (Mentions *T. expansa* and *T. denticulata*; *T. capræ* and *T. fimbriata* as doubtful species.)

11. ———. The Internal Parasites of our Domesticated Animals. London, 1873.
    (*T. expansa*, pp. 20-22; *T. denticulata*, p. 22.)

12. CREPLIN, F. C. H. Endozoologische Beiträge:1. Über *Tænia denticulata* Rud. und *T. expansa* Rud.; Weigemann's Archiv für Naturgeschichte Jarg. viii, 1842. I, pp. 315-327.

13. CURTICE, C. Tapeworm disease of the sheep of the western plains; Annual Report of the Bureau of Animal Industry. Washington, D. C., 1887-'88, pp. 167-184, pl. i-ii, *T. fimbriata*. Review by M. Braun in Centralblatt f. Bacteriologie und Parasitenkunde, 1890, viii, p. 732.

14. ———. The Animal Parasites of Sheep. Special Report of the Bureau of Animal Industry, U. S. Department of Agriculture. Washington, D. C., 1890.

 (*Tænia alba*, p. 89; *T. Benedeni*, p. 89; *T. centripunctata*, p. 89; *T. fimbriata*, pp. 89–112, pls. xii and xiii; *T. Giardi*, p. 89; *T. globipunctata*, p. 89; *T. ovilla*, p. 89; *T. oripunctata*, p. 89; *T. Vogti*, p. 89; *T. expansa*, pp. 113–126, pls. xiv and xv.)

* 15. ———. Parasites, being a list of those infesting the domesticated animals and man in the United States; Journal of Comparative Medicine and Veterinary Archives, 1892, pp. 223–236.

 (*T. expansa*, pp. 225, 227; *T. denticulata*, p. 227, is not *T. denticulata*, C. W. S.)

16. DAVAINE, C. Traité des entozoaires et des maladies vermineuse. Paris, 1877.

 (P. liii and p. 235 *Tænia expansa* Rud. and *T. denticulata* Rud.)

[17] 2. Deutsche Nordpolarfahrt I. 2. Abtheilung, 1874, p. 686.

 (*Tænia expansa*.)

18. DEWITZ, JOH. Die Eingeweidewürmer der Haussäugethiere. Berlin, 1892.

 (*Tænia expansa* Rud., p. 77–81, figs. 51 and 52.)

19. DIESING, K. M.—Tropisürös ünd Thysanosoma, zwey neñe Gattöngen von Binnenwürmern; Med. Jahrb. d. öst. Staat., n. Folge, vii, 105–111. Tab. iii et anat.

20. ———. Systema Helminthum, vol. i. Vindobonæ, 1850.

 (*T. expansa*, p. 497; *T. denticulata*, p. 498; *T. fimbriata*, p. 501; *Tænia capræ*, p. 552.)

21. ———. Zwanzig Arten von Cephalocotyleen; Denkschrift d. math. naturw. Classe der Kaiserlichen Akademie d. Naturwissenchaften. Wien, 1856.

 (*T. fimbriata*, pp. 32–33, taf. v, 9–15.)

22. DUJARDIN, F.—Histoire Naturelle des Helminthes. Paris, 1845.

 (*Tænia expansa*, p. 577; *T. denticulata*, p. 578.)

23. FAVILLE, GEO. C.—Report of the Veterinary Department of the Colorado State Agricultural College, Jan., 1885.

 (*Tænia expansa* Rud., *ex parte=Thysanosoma actinioides*.)

24. GIEBEL, C.—Die im zoolog. Museum der Universität Halle aufgestellten Eingeweidewürmer nebst Beobachtungen über dieselben; Zeitsch. f. d. gesammt. Naturw., 1866.

 (p. 259, *T. expansa ;* pp. 259–260, *T. crucigera ;* p. 265, *T. capræ*.)

25. GMELIN.—Linné's Systema Naturæ, p. 3074; No. 55. 1789–1790.

 (*Tænia ovina=T. expansa* (after Rud.).

26. GOEZE, J. A. E.—Versuch einer Naturgeschichte der Eingeweidewürmer thierischer Körper. Leipzig, 1782, 44 taf.

 (*T. ovina*, 369, taf. xxv, iii, 1–12.)

27. GURLT.—Lehrbuch der pathologischen Anatomie der Haussäugethiere, 1831, Bd. I.

 (p. 381, pl. x, figs. 1, 2, *Tænia expansa ;* plate x, figs. 3, 4, *Tænia denticulata*.)

[28] Hertwigs Magazine der Thierheilkunde, iv Jahrg., 2 Heft.

[29] HAVEMANN.—

[30] HUMBOLD.—Berliner thierärztl Wochenschrift, 1889.

 (*T. denticulata*, 45 meters long ! !)

31. HUTCHINSON, J.—On Tapeworms and other Parasites in Sheep and Rabbits; Archiv surg. London, 1891-'92, pp. 155–156, 172–175.

[32] KRABBE.—

33. LEUKART, R.—Bandwürmer; Koch's Encyklopädie der gesammten Thierheilkunde. Wien und Leipzig, 1885, pp. 361–404.

 (*T. expansa*, pp. 398–399; *T. denticulata*, pp. 399, 400; *T. alba*, p. 400.)

34. ———. Die Parasiten des Menschen, 2 Aufl. I, p. 353, 1879–1886.

35. v. LINSTOW, O. Compendium der Helminthologie, Hanover, 1878. Nachtrag, 1889.

[36.] MAYER. Froriep's Neue Notizen, i, p. 107.

(*Tænia denticulata*=*T. expansa*, after Creplin.)

[37.] ———. Analecta für vergl. Anat. 2. Samml. 70, figs. 4, 5 (de organ genit.)

[*T. denticulata*=*T. expansa*, after Crep. and Dies.]

38. McMURRICH, J. P. Zoölogical Report; T. G. Grenside, Veterinary Report, pp. 200–203. Ontario School of Agriculture, 9th Annual Report, 1884 (pp. 174–178, 7 Figs., *T. expansa*=*M. trigonophora*). (Review by V. Linstow, Archiv fur Naturgeschichte, 1888.)

39. MONIEZ, R. Note sur deux espèces nouvelles de Tænias inermes. *T. Vogti* et *T. Benedeni;* Bulletin Scientifique du Départment du Nord, 1879, sér. ii, T. ii, p. 163–164.

40. ———. Sur le *Tænia Giardi* et sur quelques espèces du groupe des inermes; Comptes Rendus de l'Academic des Sciences, lxxxviii, Mai 26, 1879, pp. 1094–1096.

41. ———. Mémoires sur les Cestodes, Paris, 1881.

(*Tænia expansa*, pp. 15–20, pl. i, figs. 38–58; pl. ii, fig. 1. Embryology.)

42. ———. Sur quelques types de Cestodes; Comptes Rendus de l'Academie des Sciences. Mar. 6, 1882, p. 662. (*T. Giardi*).

43. ———. Les Parasites de l'Homme. Paris, 1889, pp. 123, 124, (*T. Giardi*).

44. ———. Notes sur les Helminthes; Revue Biologique du Nord de la France, Tome iv, 1891.

(Extrait: V. *Moniezia ovilla*, pp. 11–14; VI. 1. *M. Benedeni*, pp. 14–16; 2. *M. Neumanni* sp. n., pp. 16–17; 3. *M. nullicollis* sp. n., p. 17; 4. *M. denticulata et M. expansa*, pp. 18–21; 5. *M. alba* var. *dubia et M. ovilla* var. *macilenta*, pp. 21–23; VII. Tableau synoptique. pp. 23–24.)

45. NEUMANN, L. G. Traité des Maladies parasitaires non microbiennes des Animaux domestiques, Paris, 1888.

(*Tænia aculeata*, p. 383; *T. alba*, pp. 378, 382; *T. Benedeni*, p. 383; *T. capra*, p. 389; *T. centripunctata*, p. 383; *T. denticulata*, p. 377; *T. expansa*, pp. 377, 382, figs. 164–166; *T. Giardi*, p. 383; *T. globipunctata*, p. 384; *T. ovilla*, p. 383; *T. ovipunctata*, p. 384; *T. Vogti*, p. 383.)

46. ———. Observations sur les Ténias du Mouton; Compt. Rend. de la Société d'Histoire Naturelle de Toulouse. Séance, 10 fév. 1891, 4 pages.

47. ———. Sur la place du *Tænia ovilla* Riv. dans la classification; Compt. Rend. de la Société d'Histoire Naturelle de Toulouse. Séance, 2 mars, 1892, 3 pages.

48. ———. Traité des Maladies parasitaires non-microbiennes des animaux domestiques, Paris, 1892, Deuxième édition.

(*Tænia expansa*, pp. 402, 407, 417, Figs. 182, 184–186; *T. denticulata*, p. 401, Fig. 183; *T. alba*, pp. 403, 408, Fig. 187; *T. Benedeni*, p. 408, Fig. 192; *T. capra*, p. 417; *T. centripunctata*, 409, Fig. 197; *T. fimbriata*, p. 408, Figs. 193–194; *T. globipunctata*, p. 410; *T. ovilla*, p. 408, Figs. 195–196; *T. ovipunctata*, p. 410; *T. Vogti*, p. 408.)

48A. *Neumann's* Parasites and Parasitic Diseases of Domesticated Animals. (Translation of 48 by Fleming.) London, 1892.

[49.] NORDMANN.—Lamark Anim. s. vert., 2 ed., iii, 591. *Thysanosoma actinioides.*

[50.] PARONA.

51. PERRONCITO, E. I Parassiti dell' Uomo e degli Animali Utili. Milano, 1882.

(*Tænia alba*, p. 243, figs. 103–105; *T. aculeata*, p. 246; *T. centripunctata*, p. 242; *T. expansa*, p. 239, figs. 99, 100; *T. globipunctata*, p. 240; *T. ovilla*, p. 244; *T. ovipunctata*, p. 241; *T. denticulata*, p. 240, figs. 101–102.)

[52.] ———. Di una nuova specie di Tænia (*T. alba*); Annali della R. Ac. d'Agric. d. Torino, 1879.

53. ——. Ueber eine neue Bandwurmart (*T. alba*); Archiv für Naturgeschichte, 1879, pp. 235-237, taf. xvi, figs. 1-10.

54. ——. Trattato teorico-pratico sulle malattie piu comuni degli Animali domestici dal punto di vista agricolo, commerciale ed igienico. Metodi di cura ed appendice sui migliori metodi do disinfezione dei vagoni. Torino, 1886.

(*Tænia expansa* R., pp. 231-232; *T. denticulata* R., p. 232, Fig. 78 (after Gurlt), Figs. 80-81 (not *denticulata*); *T. alba* P., pp. 233-234, Figs. 79, 82; *T. globipunctata*, Riv., p. 235; *T. ovipunctata* Riv., p. 235; *T. centripunctata* Riv., pp. 235-236; *T. ovilla* Riv., pp. 236-237; *T. aculeata* Riv., pp. 237-238.)

55. RAILLIET, A. Éléments de Zoologie Médicale et Agricole, Paris, 1886.

(*Tænia expansa*, p. 257, figs. 151-153; *T. denticulata*, p. 260; *T. alba*, p. 260; *T. globipunctata*, pp. 260, 261; *T. ovipunctata*, p. 261; *T. centripunctata*, p. 261; *T. aculeata*, p. 261; *T. Benedeni*, p. 261; *T. Vogti*, p. 261; *T. capra*, p. 262; *T. Giardi*, p. 264.)

55A. ——. Second edition, 1893 is in press.

[56.] "RAULIN. Traité des maladies occasionées par les promtes et frequentes variations de l'air, Paris, 1752, p. 444, a tapeworm 26 feet long from a 3 months lamb."

[57.] RIVOLTA, S. Sopra alcune specie di Tenie della Pecora, Pisa, 1874. (*T. centripunctata*, *T. ovipunctata*, *T. globipunctata*.)

58. ——. Di una nuova specie di Tænia nelle pecora (*Tænia ovilla*); Giornalo di Anat. Fisiol. et Patol. degli Animali, Pisa, 1878, pp. 302-308, figs. 1-3.

[59.] ——. Studi fatti nel Gabin. di Anat. patalog. di Pisa, 1879, p. 79, 3 figs. *T. ovilla*.

60. ROBERTSON, W. Tænia in lambs; The Veterinarian, London, 1875, vol. xlviii, p. 80.

61. RUDOLPHI, C. A. Entozoorum sive vermium intestinalium Historia naturalis. Amstelædami, 1809, vol. ii, p. 2.

(*Tænia expansa*, pp. 77-79; *T. denticulata*, pp. 79-80; *Tænia capreoli* and *T. capræ*, pp. 200, 201, [Species dubiæ.])

62. ——. Entozoorum Synopsis cui accedunt Mantissa duplex et indices. Vindobonnæ 1819.

(*Taenia capreoli*, p. 144; *T. denticulata*, p. 145; *T. expansa*, pp. 144, 487. *T. capræ* p. 170.)

63. SCHRANK, F. V. P. Verzeichniss der bisher hinlänglich bekannten Eingeweidewürmer, nebst einer Abhandlung über ihre Auverwandschaften. München, 1788, p. 80.

64. STEEL, J. H. Diseases of the Ox. London, 1881. (*Tænia expansa*, p. 294.)

65. STILES, C. W. Notes on Parasites—8: A check list of the animal parasites of cattle; Journal of Comparative Medicine and Veterinary Archives, New York, 1891, p. 346.

(*Tænia expansa*, *T. denticulata*, *T. alba*, *T. centripunctata*, *T. globipunctata*.)

66. ——. Notes sur les Parasites—13: Sur le *Tænia Giardi* (Riv.) Moniez; Comp. rend. de la Société de Biologie. Paris, 1892, pp. 664-665.

67. ——. Notes sur les Parasites—14: Sur le *Tænia expansa* Rud.; Compt. Rend. de la Société de Biologie. Paris, 1892, pp. 665-666.

68. ——. Bemerkungen über Parasiten—17: Ueber die topographische Anatomie des Gefässystems in der Familie Tæniadæ; Centralblatt für Bakteriologie und Parasitenkunde, 1893, bd. xiii, pp. 457-465, figs. 1-12.

(*M. planissima*, *M. Benedeni*, *M. Neumanni*, *M. expansa*, *M. oblongiceps*, *M. trigonophora*, *M. denticulata*, *M. alba*, *Thysanosoma actinioides*, *Th. Giardi*, *Tænia centripunctata*, *T. globipunctata*.)

69. —— and HASSALL, ALBERT. A Revision of the Adult Cestodes of Cattle Sheep, and Allied Animals. Bulletin 4, Bureau of Animal Industry, U. S. Department of Agriculture, Washington, D. C., 1893 (the present paper). Pages 1-134, 16 plates.

[70.] TABL. ENCYCL. t 45. (Figs. 1–12. (i. c. Goeze). *Tænia ovina=T. expansa* (after Rud.).

71. VERRILL, A. E. Parasites of Domestic Animals; Report of the Connecticut Board of Agriculture, 1870.
(*Tænia expansa*, pp. 65, 100; *T. denticulata*, p. 65.)

72. VILLAR, Sydney. The Veterinarian, Oct., 1886.

[73.] ZEDER, J. G. H. Anleitung zur Naturgeschichte der Eingeweidewürmer. Bamberg, 1803.
(*Halysis ovina=T. expansa*, after Rud.)

74. ZSCHOKKE, F. Recherches sur la Structure Anatomique et Histologique des Cestodes. Genève, 1890. (*Tænia expansa*, pp. 93–114, figs. 31–38.)

75. ZÜRN, F. A. Die thierischen Parasiten anf und in dem Körper unserer Haussäugetiere, Wiemar, 1882.
(*Tænia expansa*, p. 189, taf. iii, figs. 39–41; *T. denticulata*, p. 197; *T. alba*, p. 197; *T. ovilla*, p. 198.)

Several other references were found but we have been unable to trace the articles. S. & H.

ADDENDA.

As the manuscript of the preceding pages left the author's hands some months ago, it may be well to add a postscript covering several interesting observations made since that time. During the past six months it has been possible to examine a large number of specimens of adult cestodes of cattle and sheep, and thus to put the analytical table given on pp. 88, 89 to a very severe test. The parasites examined came from the several sources enumerated below:

(a) Thirty strobilæ collected by Dr. Melvin and assistants at the abattoirs of the Union Stock Yards of Chicago. These strobilæ were preserved in weak alcohol and did not come into my hands until they were quite macerated. Although disappointed at first because the material was not in better condition, its examination has led to some very important and rather unlooked-for results.

(b) A few segments of Rudolphi's original *Tænia capræ*, kindly forwarded by Geheimrath Karl Möbius.

(c) About 100 strobilæ of *Moniezia* from cattle and sheep collected by Dr. Hassall at the Washington (D. C.) abattoir, and fixed according to the method described on pp. 13, 14.

(d) Several jars of tapeworms from sheep belonging to the Leidy collection, University of Pennsylvania. The specimens are all members of the genus *Moniezia*, but are very much macerated.

(e) Several strobilæ collected by Dr. Theobald Smith.

(f) About 100 strobilæ collected by myself in August, from cattle at the abattoirs of Chicago, and preserved according to the method described on p. 13.

The specimens collected by Hassall and myself were very readily determined as belonging to the species *M. planissima*, *M. expansa*, and *M. trigonophora*. Several strobilæ bear a very close resemblance to *M. Neumanni*, but I am not willing at present to state definitely that they belong to that species. A study of the forms not " fixed" before being placed in alcohol emphasized the importance of not relying upon external form in making the specific determinations—a point to which attention has already been drawn in the text—for it is the easiest thing in the world to determine an *M. expansa* as *M. planissima*, or *vice versa*. If microscopic preparations are made, however, there is not the slightest difficulty in distinguishing the *Planissima group* from the *Expansa group*. In many cases, however, it is difficult to distinguish *M. expansa* from *M. trigonophora* by examining only one or two segments, but the species are easily separated by a more careful study.

The specimens of Leidy's collection, although too macerated to be determined specifically, could easily be divided into the *Planissima group* and the *Expansa group*.

97

The results of the study confirm the position taken in the preceding text, that the form of the interproglottidal glands furnishes a basis for a natural division into the *Planissima group* and the *Expansa group*. I am led, however, to doubt the advisability of placing very much stress upon the size of the glands in the *Planissima group* as a criterion in determining the species; in several cases strobilæ were found which agreed with *M. planissima* in all respects except that the head was much larger and the linear glands extended the entire (or almost the entire) width of the segment. Preferring to be conservative, I have deemed it best for the present to look upon these specimens as the *M. planissima* rather than to create a new species for them, for in studying parasites, as well as other animals, *we must expect to find individual variation.* This variation can, for the present, be held to account for the enormous size of the linear glands. The enormous size of the heads in several specimens *might possibly* be accounted for by *individual variation* and *method of preservation*, for I have studied the live heads a number of times, and find that, in their contraction, they can change in size to a considerable degree.

An interesting fact was developed in comparing unfixed and fixed material with Perroncito's types of *M. alba*. Specimens of *M. planissima* when fixed with sublimate bear absolutely no resemblance to Perroncito's *M. alba* (the type specimens of *T. alba* were evidently preserved in alcohol without first being treated with sublimate); specimens of *M. planissima* which have not been fixed in sublimate, but have remained in weak (50 per cent) alcohol for some time, differ from specimens which have been fixed in sublimate to a marked degree. In fact, when I first examined the specimens macroscopically I was entirely at a loss to know to which species they might possibly belong. A microscopic examination of a number of slides, however, left no doubt that they were typical specimens of *M. planissima*, and that any difference in the form of the segments or the general appearance of the strobila should be accounted for by considering the different methods used in preserving the specimens. Next came an interesting, and to my mind an important observation, i. e., that a few of the more poorly preserved of the alcohol specimens bore a remarkable macroscopic resemblance to Perroncito's type of *T. alba*. In fact, in the case of one specimen the resemblance was so striking that when I compared it side by side with one of Perroncito's specimens I was unable to tell which was the Italian and which the American form without comparing the labels. Having called attention to this remarkable macroscopic resemblance, it is in place to compare the microscopic slides. In no case have I found a single slide, in the many preparations made from three of Perroncito's types, in which there was the slightest trace of interproglottidal glands on the other hand, I have never failed to find these linear glands in the segments of *M. planissima* which were treated with sublimate; in the vast majority of cases those nonfixed specimens of American tape worms which did not belong to the *Expansa group* have possessed linear interproglottidal glands. Most of these are undoubtedly our *M. planissima;* on the other specimens I prefer to reserve judgment for the present. The same character was present in French specimens sent to me by Railliet and Neumann. Now, curiously enough, seven nonfixed fragments of American forms fail to show any signs of interproglottidal glands. These specimens are poorly preserved, but not badly macerated; under the microscope they can not be distinguished from *M. alba.* I should not think of uniting them with *M. planissima,* on account of the absence of the linear interproglottidal glands; if such glands were present they would probably have been brought out by staining, for we have frequently found these glands in fragments which were more macerated than those now under discussion. If pressed to a determination of these fragments, I should accordingly label them *M. alba,* although rather unwillingly, because they are only fragments and are the only specimens of *M. alba* (?) ever collected in America.

I might add, further, that two of Perroncito's types agree perfectly with each other when examined macroscopically, while the third, as well as Neumann's specimens, would probably not be united with *M. alba* by any worker except after a careful microscopic study.

To recapitulate, in a few words, we may conclude that—

(1) Specimens of a species fixed in sublimate may differ so in macroscopic appearance from specimens of the same species simply preserved in alcohol that scarcely any helminthologist would hesitate to recognize them as belonging to different species.

By using different methods of preservation and different degrees of maceration it has been found possible to bring about such a variation in external form of different specimens of *M. planissima* and *M. expansa,* or of different portions of the strobila of a single specimen that almost any species (if established on external form alone) can be made to order.

(2) It is entirely unsafe to determine a given specimen of *Moniezia* as *M. planissima, M. expansa,* or in fact as almost any other species without first examining several stained segments. Segments with well-developed testicles and ovaries are the best for this purpose.

(3) As specimens macerate, the linear interproglottidal glands of *M. planissima* stain less distinctly.

(4) In Perroncito's type of *M. alba* there is no trace of any interproglottidal glands; these specimens were evidently not fixed in sublimate before being placed in alcohol, but they are not badly macerated.

(5) One specimen of an undoubted *M. planissima* resembles one specimen of Perroncito's type so very closely (macroscopically) that no zoölogist would hesitate to pronounce the two specifically identical, were it not that linear glands are present in the former and absent in the latter.

(6) The specimen of *M. alba* referred to (5) bears no macroscopic resemblance to Perroncito's other two types, nor to his figures of this species, but on account of the absence of the glands it must be united with *M. alba*.

Upon reading the above, some of my European colleagues will undoubtedly conclude that *M. planissima* and *M. alba* are specifically identical, that the absence of the linear gland is to be accounted for by maceration, and that we have been too hasty in creating *M. planissima* as a new species. Such a conclusion would, however, be entirely unjustified unless the person based his opinion on a careful study of the types and a large series of typical specimens. If anyone is able to find linear interproglottidal glands in Perroncito's types and to prove that Perroncito's types contain but one species, then he would be perfectly justified in uniting *M. planissima* and *M. alba* as one species, and no one would be more willing to accept his results than Hassall and myself. I have not been able to find these glands in *M. alba*, but have found them in every specimen of tapeworm of the genus *Moniezia* except those of the *Expansa group*, the few segments of *M. denticulata* now at my disposal, in Perroncito's and Neumann's specimens of *M. alba*, and the few fragments mentioned above as *M. alba* (?).

This Bureau has already sent to a number of zoölogists typical specimens (stained slides and alcohol specimens) of the forms described in this paper, and it can safely be left to European specialists to decide whether we were justified in creating the species *M. planissima*, and it is needless to add that the Bureau types and preparations are always at the disposal of specialists in helminthology. I feel that we have solved the question of the classification of this genus so far as it applies to the American forms and to the limited amount of European specimens at our disposal. If we have made any errors in judging the European forms they can best be corrected by our European colleagues. Looking to the possibility of an European revision of this genus in the future I should like to call particular attention to the following points:

(1) Are the scarcely visible linear glands in *M. Benedeni*, mentioned in this paper, in reality interproglottidal glands or have I here mistaken some of the more intensely stained subcuticular cells for glandular cells? This point can be determined only by an examination of properly preserved, freshly collected material. If I have here made an error of interpretation, the species should be placed in the "*Denticulata group*."

(2) Are interproglottidal glands present in freshly collected and properly preserved material of *M. denticulata?*

(3) Can the species *Stilesia centripunctata* remain in the same genus with *S. globipunctata?* This certainly can not be so unless we find more points of resemblance between the two forms than we know at present.

(4) Any classification of any genus of tapeworms, based upon external form alone, can not claim the attention of specialists. Anatomy must form the basis of our future revisions.

(5) Individual variation must be taken into consideration in the study of parasites just as well as in the study of other animals. Slight, or, in many cases, even considerable, difference in the size of the head, form of the segments, etc., can often be explained as due to different methods of preserving.

Moniezia capræ sp. inq.

The segments of Rudolphi's *T. capræ*, now in my possession, warrant but one statement, *i. e.*, that the species is evidently a true *Moniezia*.

The proglottids are so crowded with ova that one would scarcely expect to distinguish the interproglottidal glands, even if they are present. Experience in comparing the external form of specimens has taught me so many lessons that I would not even dare to speculate as to the possible affinities of this specimen with the other species of *Moniezia*.

Thysanosoma actiniodes.

This species has been found in South Africa (personal conversation with Dr. S. Wiltshire).　　　　　　　　　　　　　　　　C. W. S.

OCTOBER 28, 1893.

INDEX TO SPECIFIC NAMES.

The names in italics are synonyms. In the case of such forms as *T. marmotæ, T. pectinata*, etc., whose generic position is somewhat uncertain, I have retained them provisionally in the collective genus *Tænia*, pending the establishment of new genera. Private advices from Prof. Railliet inform me that he intends to establish several new genera in the second edition of his *Traité de Zoologie médicale et agricole* (now in press). According to Prof. Railliet's letters he has adopted the MS generic name *Andrya* for *T. marmotæ* and *Ctenotænia* and *Rhopalocephala* for Riehm's species. As Railliet's second edition has not yet reached this country, at the time of the final proof-reading and indexing of this bulletin, I am not in a position to state positively whether Railliet has retained in his work the manuscript names given above, and the reader must, therefore, be referred to Railliet's publication for the generic names to be employed in connection with the species in question.   C. W. S.

	Page.
*Alyselminthus denticulatus**	42
*expansus**	26
Anoplocephala Vogti	83
Dipylidium latissimum	53, 71
Leuckarti	53, 71
pectinatum	53, 71
*Halysis ovina**	26
Linguatula armillata	12
Moniezia alba	47, 98
var. dubia	47
Benedeni	22
capræ sp. inq	101
expansa	26
denticulata	42
festiva	53
fimbriata	53, 55
Goezei	53, 71
Leuckarti	53, 71
marmotæ	53, 71
Neumanni	25, 53
nullicollis	53, 83
oblongiceps	35
orilla	53, 59
var. macilenta	59
pectinata	53, 71
planissima	15, 98
trigonophora	37
Pentastomum proboscideum	12
Porocephalus annulosus	12
armillatus	12
constrictus	12
crotali	12
megastomus	12
moniliformis	12
oxycephalus	12
polyzonus	12
subuliferus	12
Stilesia centripunctata	79

	Page.
Stilesia globipunctata	73
Strongylus contortus	37
Tænia aculeata	59
alba	47, 53
Benedeni	22, 37, 53
capræ	86
capræa	86
capreoli	86
caprina	86
centripuncta'a	79
centripunteggiata	79
crassicollis	78
crucigera	85
denticolata	42
denticulata	22, 26, 42, 53
expansa	15, 26, 37, 53, 55, 74
festiva	53
fimbriata	53, 55
Giardi	53, 59
globipunctata	73
globipunteggiata	73
hyracis	70
madagascariensis	60
marmotæ	53, 71
ovi-globipunctata	74
ovilla	53, 59, 62
ovina	26, 86
ovipunctata	73
ovipunteggiata	73
ovis arietis	62
pectinata	53
saginata	32
solium	78
vasis nutriciis distinctis	26
Vogti	83
Thysonosoma actinioides	53, 55, 101
Giardi	59

*Original article not at my disposal.

PLATE I.—*Moniezia planissima.*

Fig. 1. Adult strobila, natural size, slightly contracted.

Fig. 1a. Ripe segments, uncontracted.

Fig. 2. Head, ventral view, balsam preparation. × 17.

Fig. 2a. Head, alcohol specimen. × 17.

Fig. 2b. Head, *en face*, alcohol specimen. × 17.

Fig. 3. Segments, 10ᵐᵐ from the anterior extremity, showing the first appearance of the genital anlagen.

Fig. 4. Segments, 70ᵐᵐ from the anterior extremity, showing the pistol-shape anlagen of the genital canals.

Fig. 5. Sagittal section, 20ᶜᵐ from the anterior extremity, showing the overlapping posterior flap of each segment.

104

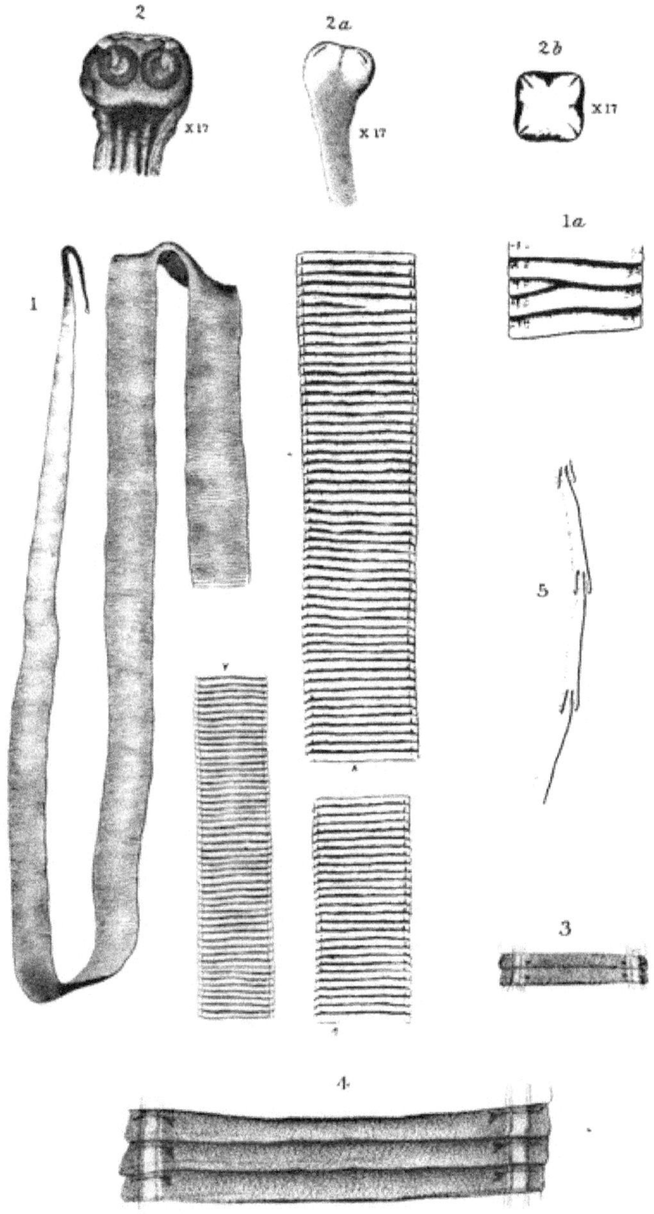

HAINES, DEL. A. HOEN & CO.

PLATE II.—*Moniezia planissima* and *M. Benedeni*.

Figs. 1-6 *M. planissima*.

Fig. 1. Diagram of the female organs, dorsal view: *r. s.*, receptaculum seminis; *ov.*, ovary; *s. g.*, shell-gland; *v. g.*, vitellogene gland; *r. dt.*, vitello-duct; *o. d., o. d'., o. d''.*, oviduct.

Fig. 2. Frontal section of the interproglottidal gland.

Fig. 3. Diagrammatic figure of a transverse section; *r.*, vagina; *c.*, cirrus; *n.*, nerve; *r. c.*, ventral canal; *d. c.*, dorsal canal.

Fig. 4. Dorsal view of segment, 200^{mm} from the anterior extremity; *i. g.*, interproglottidal gland; *t.*, testicles; *d. c.*, dorsal canal; *v. c.*, ventral canal; *n.*, nerve.

Fig. 5. Ripe segments filled by the uteri.

Fig. 6. Ovum.

Figs. 7, 8. *M. Benedeni*, drawn from one of Moniez's original specimens.

Fig. 7. Head and strobila, natural size.

Fig. 7a. Two segments, drawn to show the peculiar architecture in contraction.

Fig. 8. Scolex, *en face*. × 17.

Fig. 8a. Scolex, ventral view. × 17.

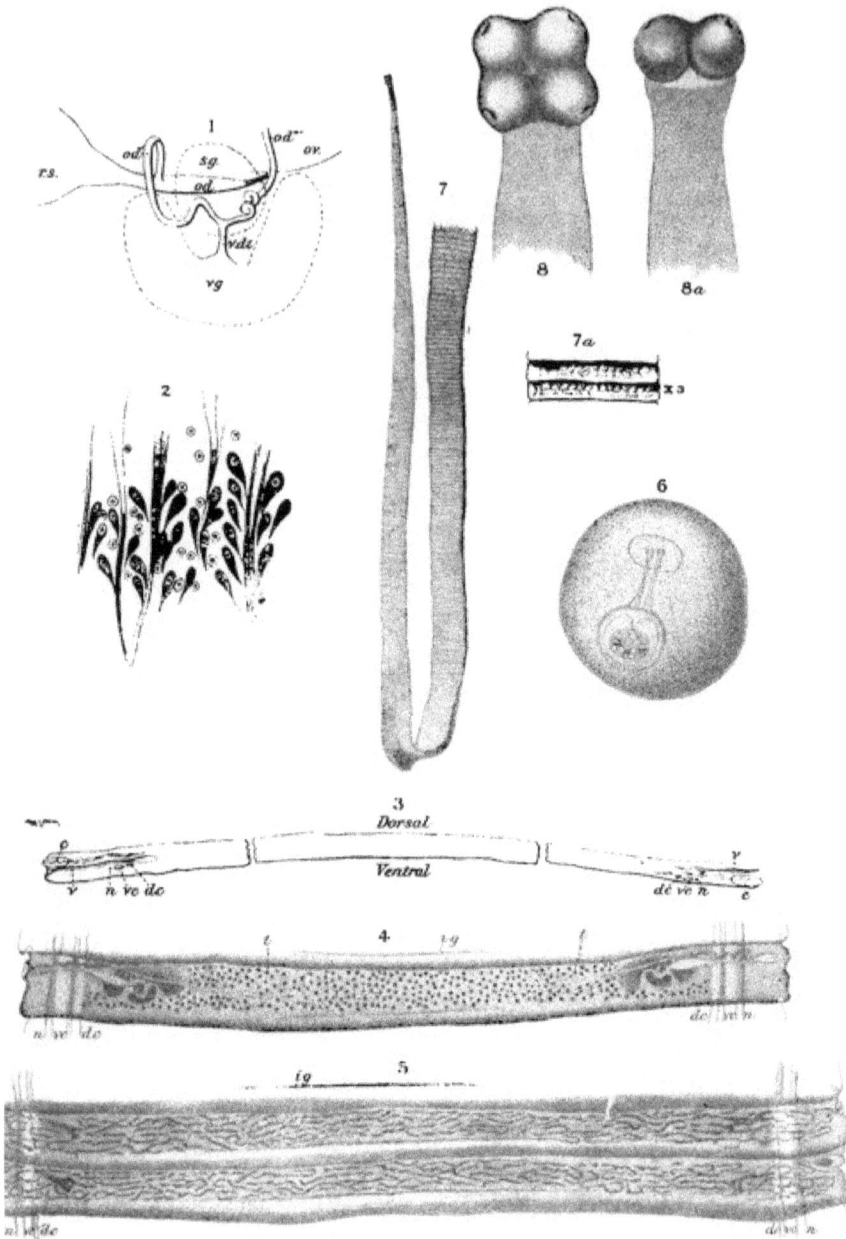

MONIEZIA PLANISSIMA   MONIEZIA BENEDENI

PLATE III.—*Moniezia planissima.*

Fig. 1. Transverse projection of the female canals, diagrammatic.

Fig. 2. Sagittal section of the interproglottidal gland.

Fig. 3. Portion of segment, showing the genital organs.

Fig. 4. Longitudinal section of cirrus and vagina, slightly diagonal.

Fig. 5. Transverse section of vagina.

Fig. 6. Transverse section of cirrus.

Fig. 7. A young egg, diagrammatic.

Fig. 8. Disk of the pyriform body.

PLATE III

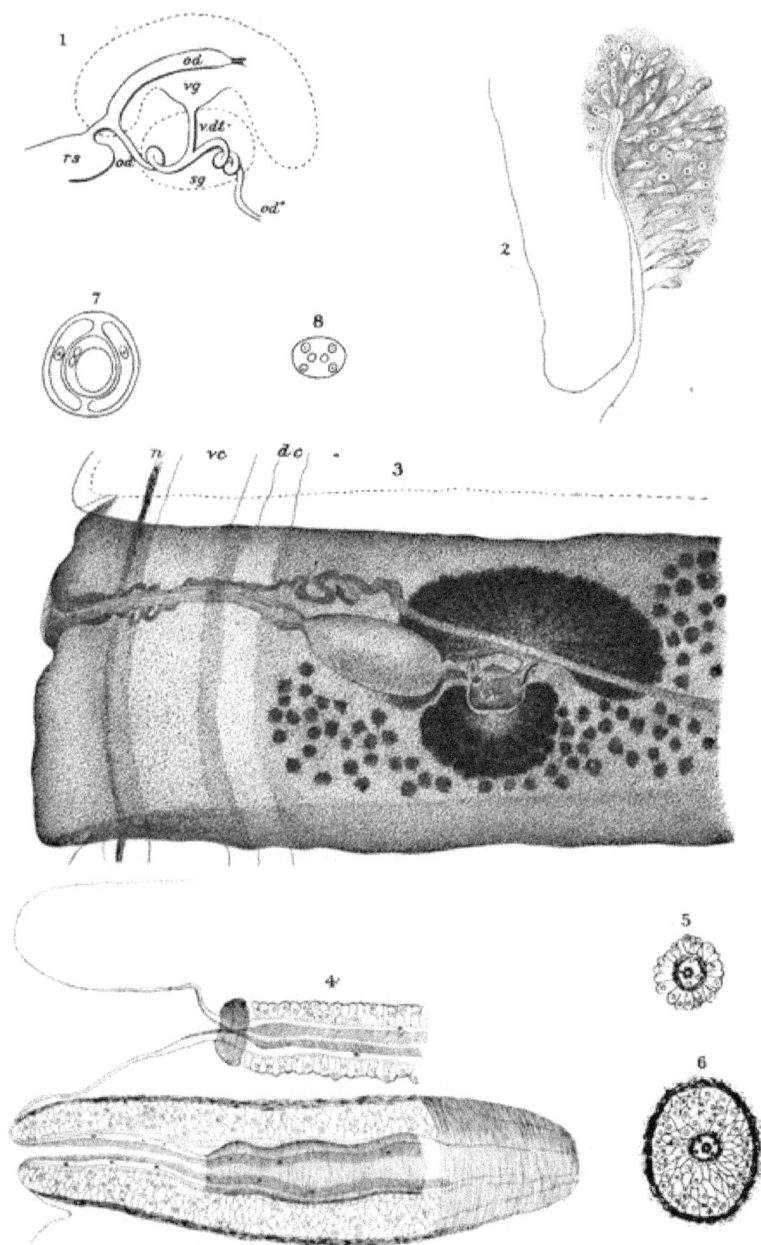

110

PLATE IV

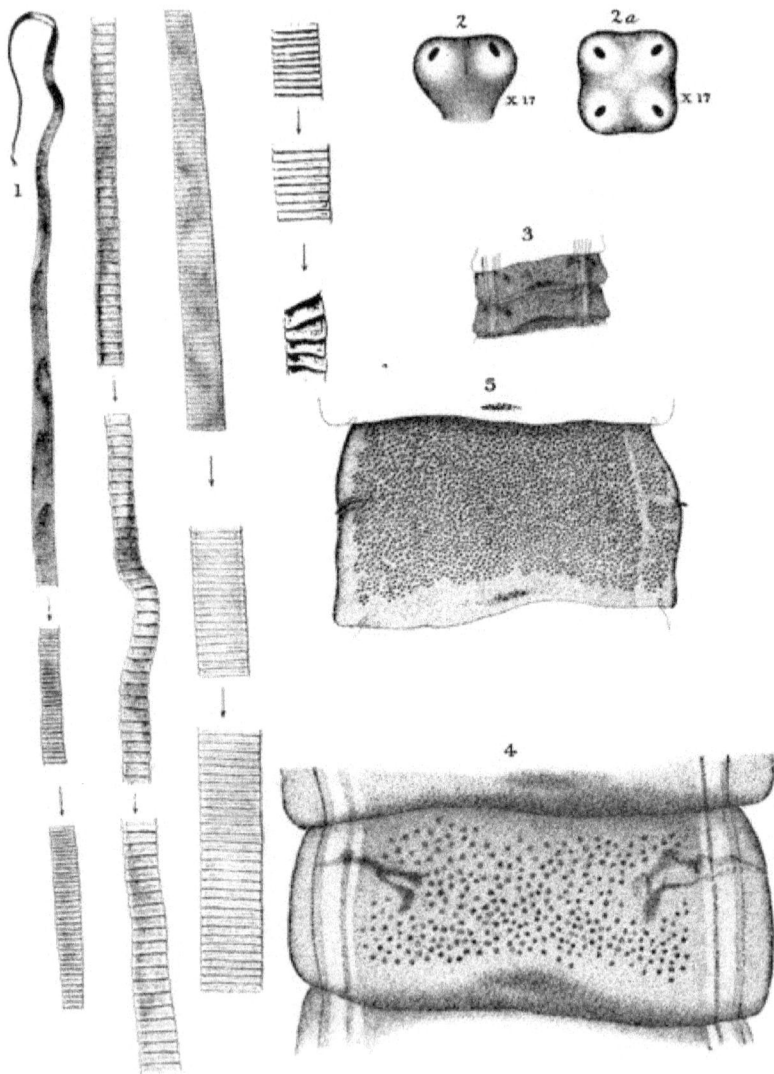

Figs. 1-3. *M. expansa*, drawn from Rudolphi's original material (1810).   Orig.

Fig. 1. Scolex, ventral view.   × 17.

Fig. 1*a*. The same, *en face*.   × 17.

Fig. 1*b*. The same, viewed diagonally.   × 17.

Fig. 2. Segments in which the uteri were not developed.

Fig. 3. Segments in which eggs were developed.

Figs. 4-6. *M. denticulata*: 4-5 drawn after Goeze's figures of Rudolphi's origin material; 6 drawn from Rudolphi's original material (orig).

Fig. 4. Strobila with head, natural size.

Fig. 5. Head.

Fig. 6. Mature segments.

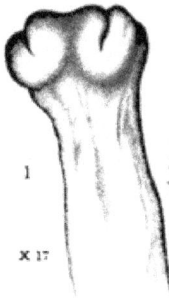

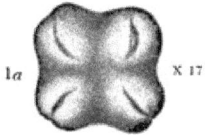

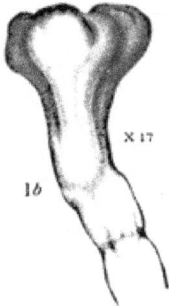

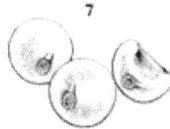

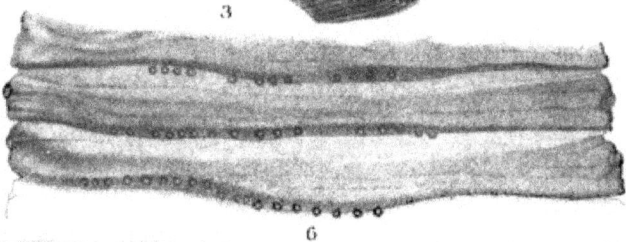

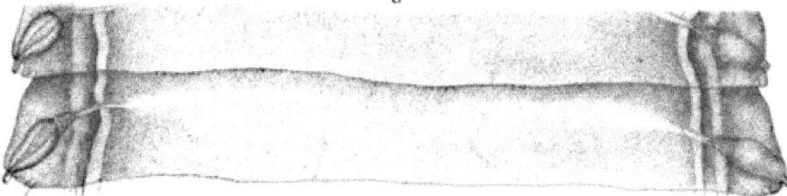

PLATE V

MONIEZIA EXPANSA   MONIEZIA DENTICULATA

RUDOLPHI'S ORIGINAL TYPES.

PLATE VI.—*Moniezia expansa.*

Fig. 1. Strobila, natural size, collected by Curtice, Washington, D. C.

Figs. 1*a*, 1*b*. Segments of Curtice's type (Animal Parasites of Sheep). × 3.

Fig. 1*c*. Segments collected at Paris. (Stiles.) × 3.

Figs. 2–2*a*. Heads collected at Washington, D. C.; very small. × 17.

Fig. 3. Segments 20ᶜᵐ from head showing interproglottidal glands, pistol-shap genital anlagen, testicles. Dorsal view.

Fig. 4. Segments showing complete anatomy except the uteri. Dorsal view.

Fig. 5. Segments from Paris (Stiles' private collection). Ventral view.

Fig. 6. Segment showing the uteri. Ventral view.

114

PLATE VI

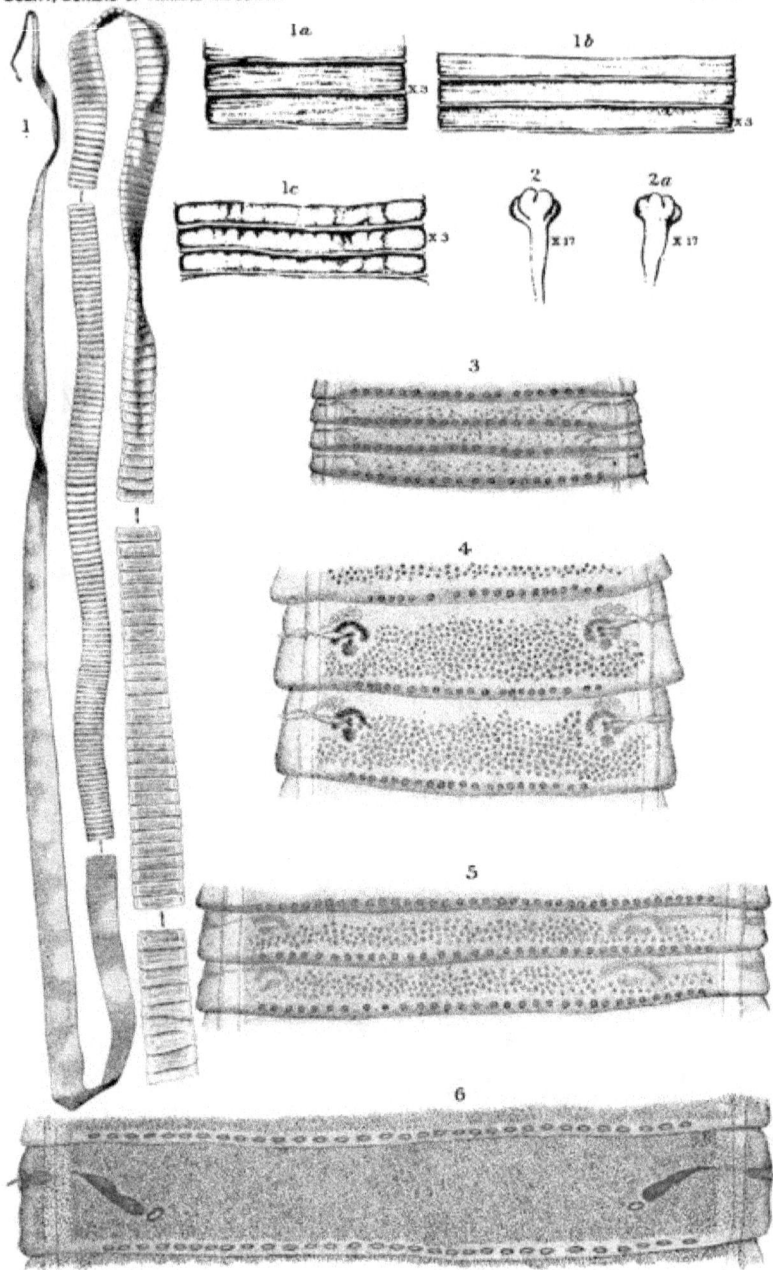

PLATE VII.—*Moniezia oblongiceps—Thysanosoma Giardi—Tœnia marmotœ.*

Figs. 1–4. *M. oblongiceps.*

Fig. 1. Strobila, natural size.

Fig. 2. Head, alcohol specimen, ventral view.

Fig. 2a. The same, *en face.*

Fig. 2b. Head, balsam preparation.

Fig. 3. Segments showing genital canals, testicles, interproglottidal glands, etc.

Fig. 4. Egg.

Fig. 5. Section of *Th. Giardi;* $g.\,p.$, genital pore; $r.\,s.$, vesicula seminalis; $v.\,d.$, convolutions of vas deferens with prostata; $v.\,d.$, vas deferens; r, vagina; $r.\,s.$, receptaculum seminis; $o.\,d.$, descending oviduct; $o.\,d'.$, ascending oviduct; $u.\,t.$, uterus; $o.\,v.$, ovarium; $v.\,g.$, vetellogene gland; $v.\,c.$, ventral canal; $d.\,c.$, dorsal canal; $t.\,c.$, transverse canal; $t.$, testes.

Fig. 6–7. *Tœnia marmotœ.*

Fig. 6. Segment showing the anatomy with the exception of the uterus, dorsal view.

Fig. 7. Segments with developed uterus.

116

PLATE VII

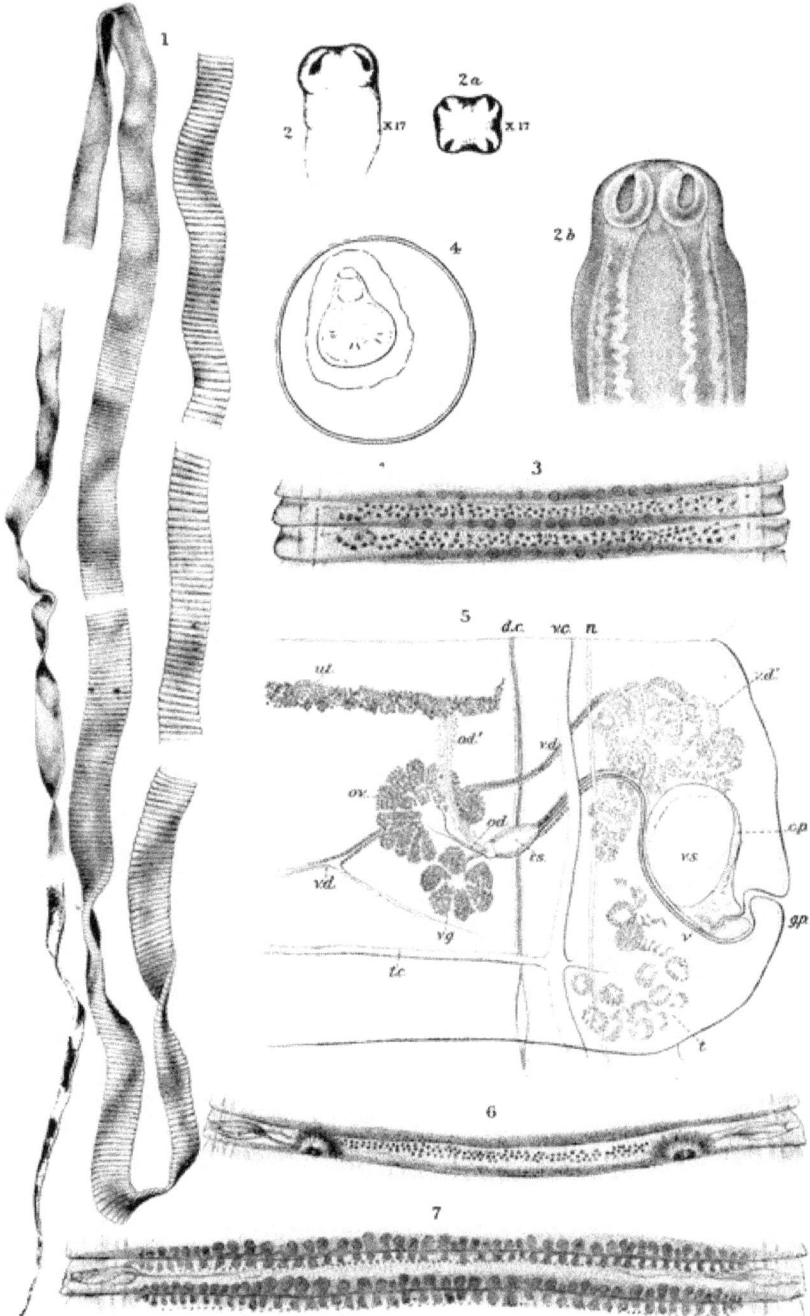

PLATE VIII.—*Moniezia trigonophora.*

Fig. 1. Strobila, natural size.

Fig. 2. Head, balsam preparation. × 17.

Fig. 2a. Head, alcohol preparation, ventral view. × 17.

Fig. 2b. The same, *en face.* × 17.

Fig. 2c. The same, lateral view. × 17.

Fig. 3. Segments 10^{cm} from anterior extremity.

Fig. 4. Segments 30^{cm} from anterior extremity.

Fig. 5. Diagram of female glands and canals, dorsal view.

118

PLATE VIII

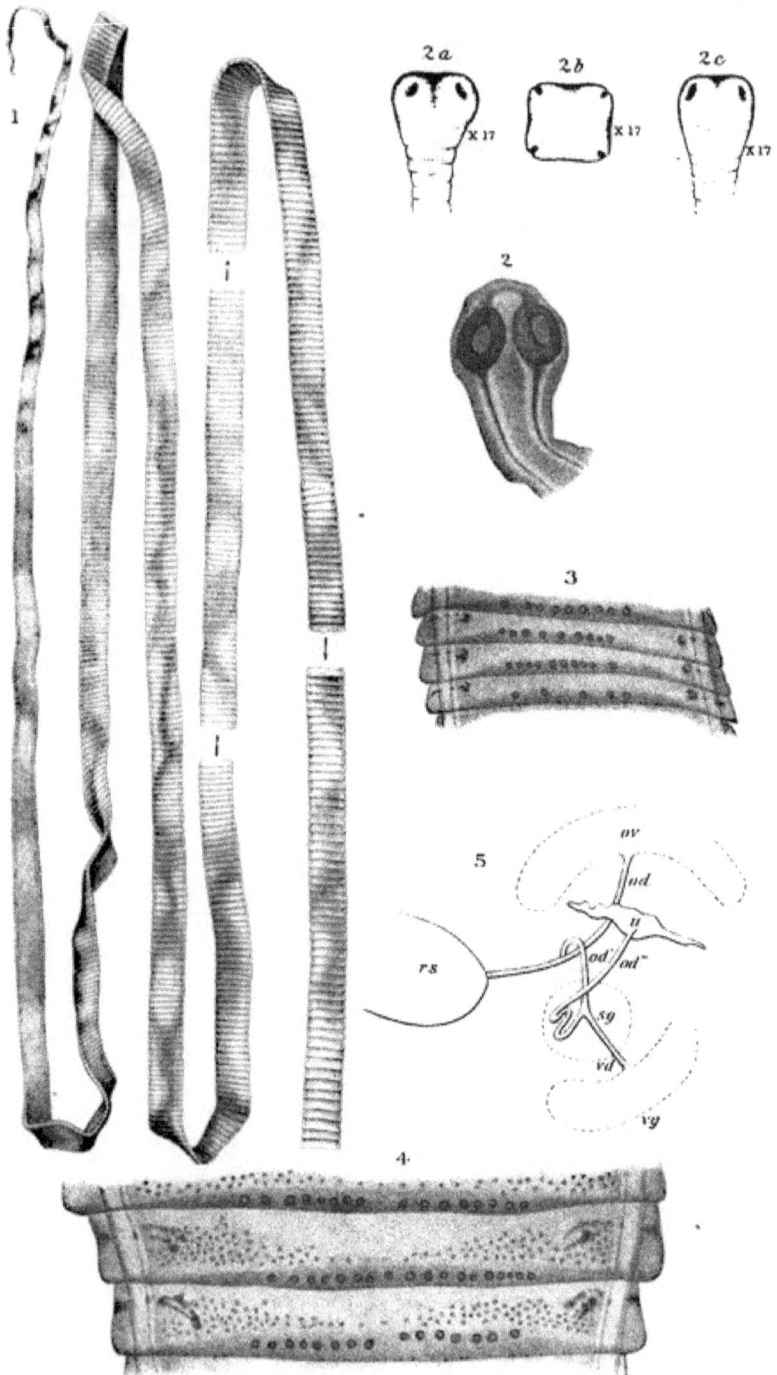

HAINES. DEL.

A. HOEN & CO.

MONIEZIA TRIGONOPHORA

PLATE IX.—*Moniezia trigonophora.*

Fig. 1. Sagittal section of a sac around which the cells of the interproglottic gland are grouped.

Fig. 2. The same, frontal section.

Fig. 3. Diagram of the female glands and canals, transverse section.

Fig. 4. Segment 60^{cm} from the anterior extremity, showing all the organs exce the uteri, which have not yet developed. Ventral view.

Fig. 5. Segments with developing uteri; a space is left vacant in the median li

Fig. 6. Segments with fully developed uteri and eggs.

120

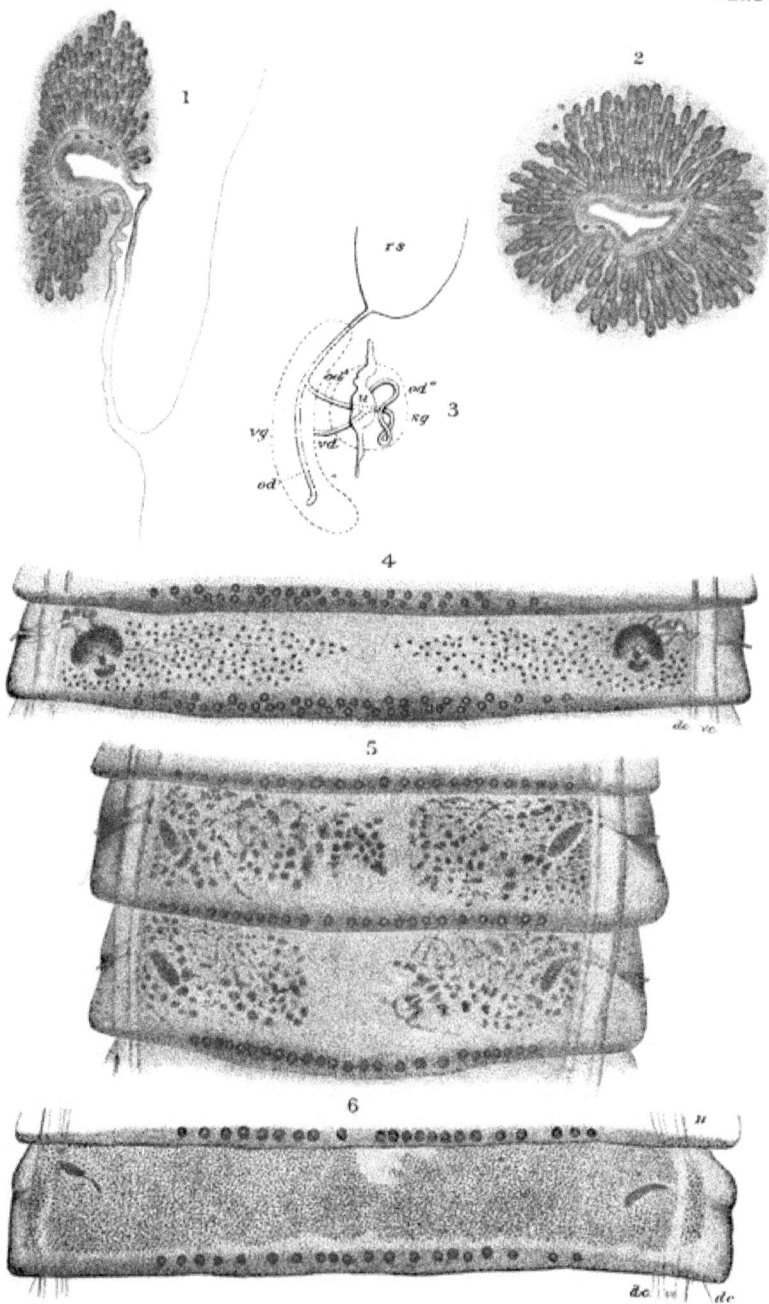

PLATE X.—*Moniezia alba.*

Figs. 1–2b, from Perroncito's original material.

Fig. 1. Strobila, natural size; a, head and anterior portion of one specimen; b–n from a specimen $1^m\,40^{cm}$ long, lacking head; b, anterior portion; c, 10^{cm} from anterior end; d 20^{cm}, e 30^{cm}, f 40^{cm}, g 50^{cm}, h 70^{cm}, i 80^{cm}, j 90^{cm}, k 100^{cm}, l 110^{cm}, m 120^{cm}, n 130^{cm}, from anterior extremity.

Fig. 2. Head, lateral view. × 17.

Fig. 2a. Head, ventral view. × 17.

Fig. 2b. Head, viewed *en face.* × 17.

Figs. 3–7 from Neumann's specimen.

Fig 3. Segments 10^{cm} from the head, 2^{mm} wide.

Fig. 4. Segments 32^{cm} from the head, 3^{mm} wide.

Fig. 5. Segments 55^{cm} from the head, 4^{mm} wide.

Fig. 6. Segments 79^{cm} from the head, 4^{mm} wide.

Fig. 7. Segments 1^m from the head, 5.5 wide by 3^{mm} long.

122

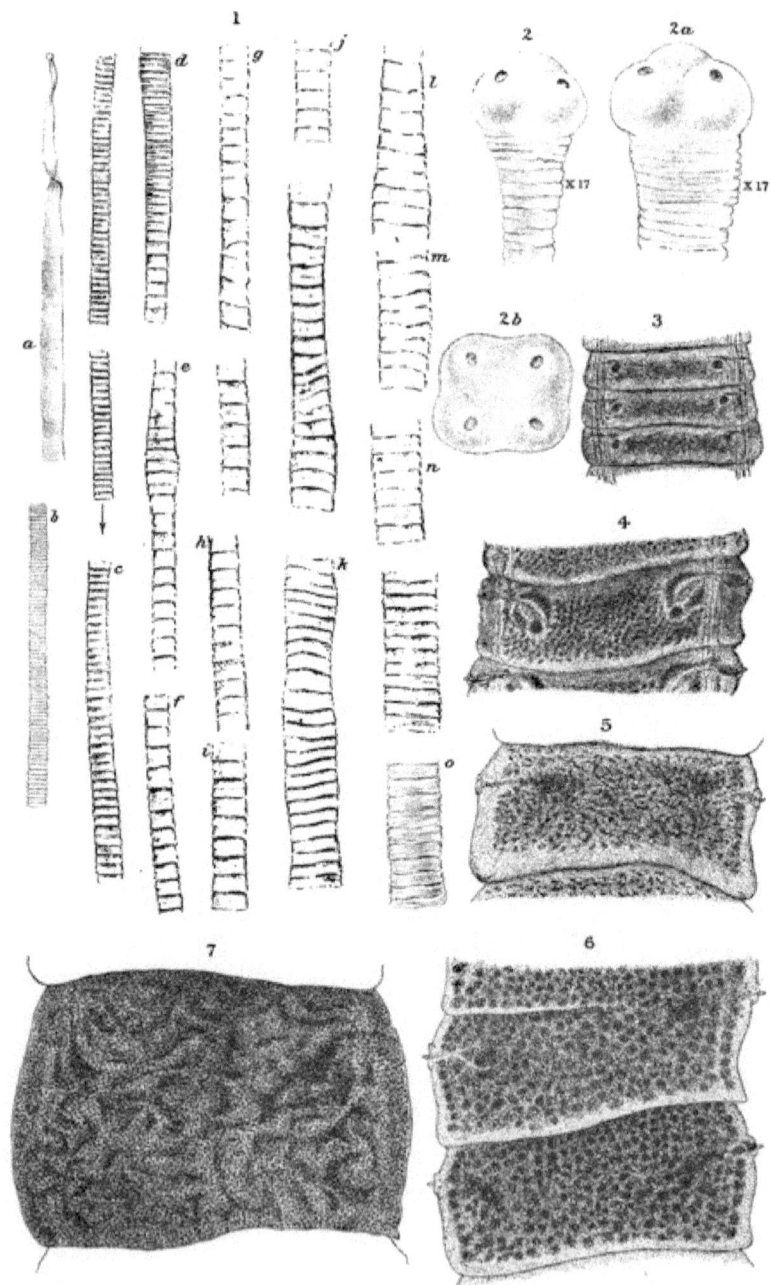

PLATE XI.—*Thysanosoma actinioides.*

Fig. 1. Strobila, natural size.
Fig. 1a. End portion of another strobila.
Fig. 2. Head, ventral view.   × 17.
Fig. 2a. The same, lateral view.   × 17.
Fig. 2b. The same, *en face*.   × 17.
Fig. 3. Segments 60ᵐᵐ from the head.
Fig. 4. Segments 100ᵐᵐ from the head, showing testicles and genital canals.
Fig. 5. Sagittal section showing two transverse canals.
Fig. 6. Transverse section showing topography of the genital and longitudinal canals.
Fig. 7. Cirrus in copulation.
Fig. 8. Segment with uterus.

124

PLATE XI

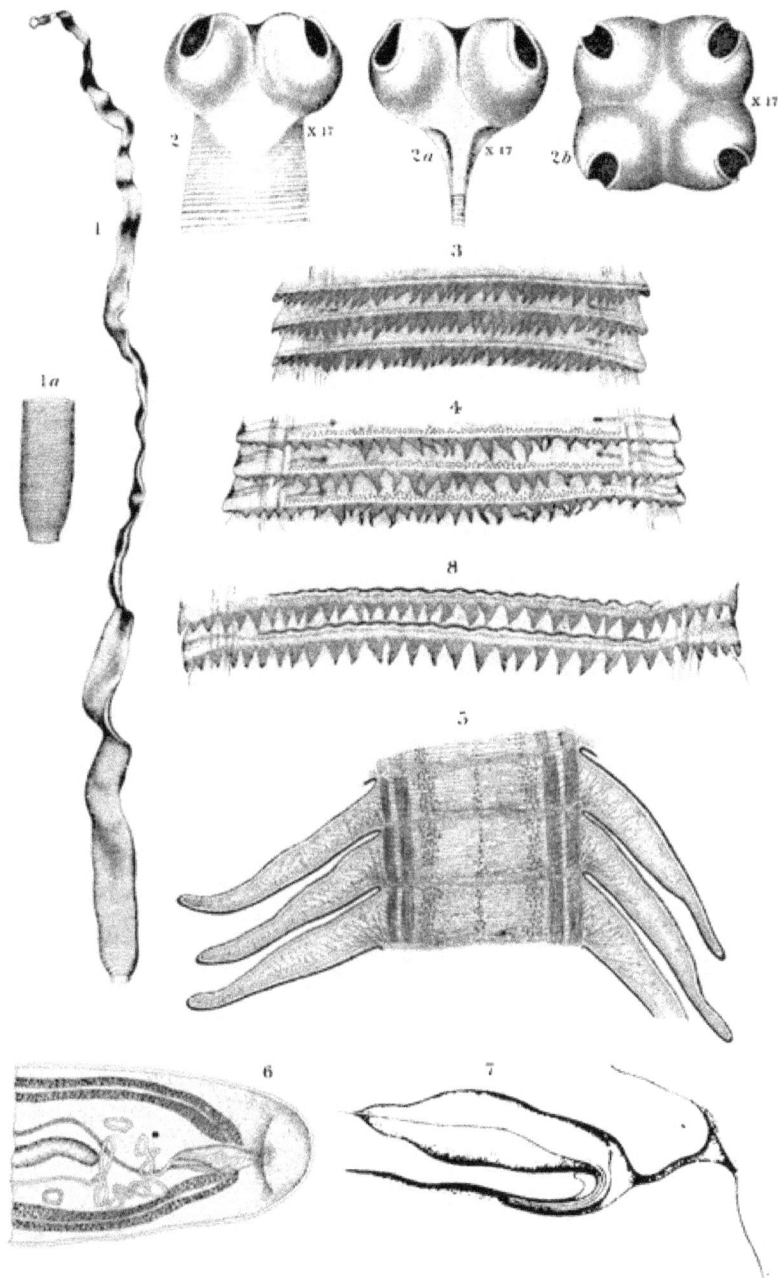

THYSANOSOMA ACTINIOIDES

PLATE XII.—*Thysanosoma Giardi.*

Fig. 1. Strobila, natural size; head drawn from a separate fragment.

Fig. 1a. Fragment, with larger segments, natural size.

Fig. 2. Head probably of *Th. Giardi*, balsam preparation. × 17.

Fig. 3. Strobila belonging to head in Fig. 2. The genital anlage is at first in the median line, then diverges towards the right and left in alternate segments.

Fig. 4. Segments with pistol-shaped anlage of the genital canals. One segment has an extra male canal. Dorsal view.

Fig. 5. Six segments with developing genital organs. Dorsal view.

Fig. 6. Genital organs well developed. Dorsal view. See Fig. 5, Plate VII.

Fig. 7. Sagittal section showing the proportion of length to thickness. T. c. transverse canal.

Fig. 8. Three segments showing the wavy appearance of the uterus.

126

PLATE XII

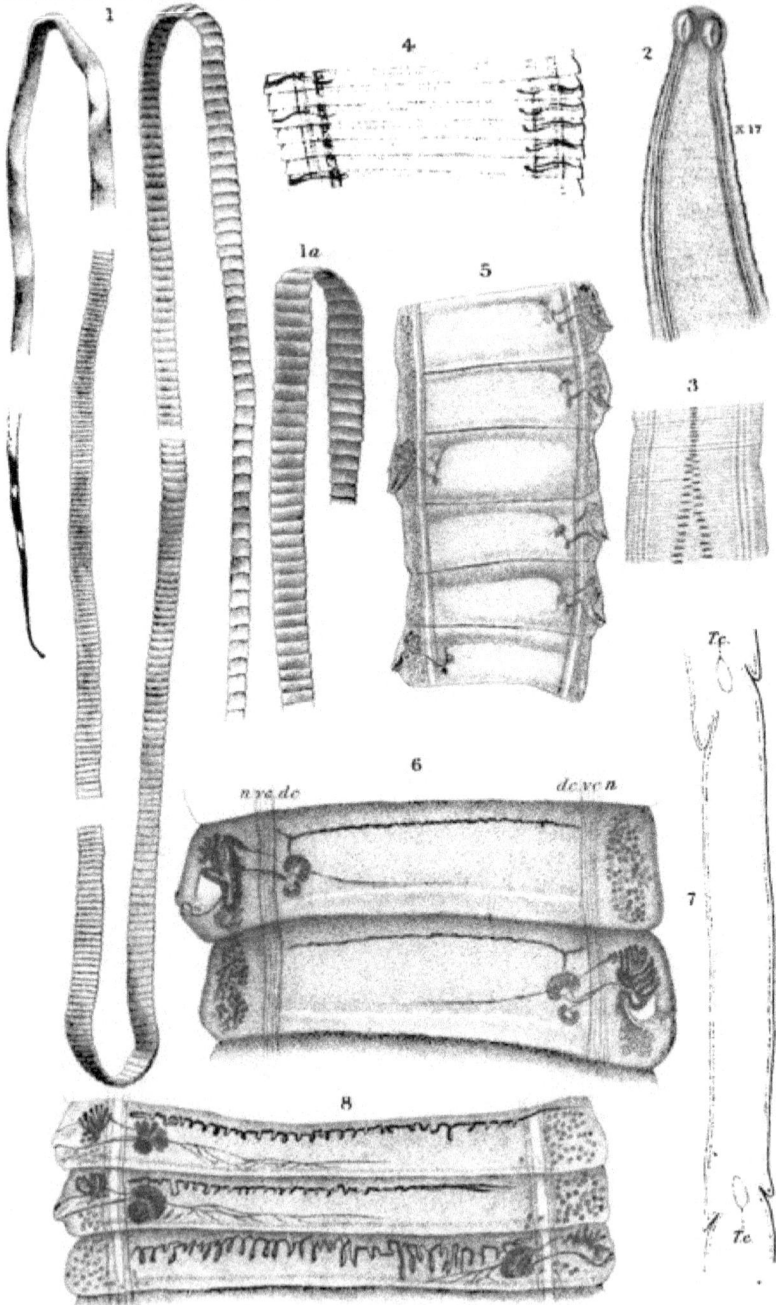

THYSANOSOMA GIARDI

PLATE XIII—*Thysanosoma Giardi.*,

Fig. 1. Three segments, ventral view. The second segment has double genita-pores.

Fig. 2. Segments with well-developed uterus. The other genital organs are atro-phying. Dorsal view.

Fig. 3. Segment with fully-developed uterus. The other genital organs are almost entirely atrophied.

Fig. 4. Egg-sacs developing.

Fig. 5. Fully developed egg-sacs.

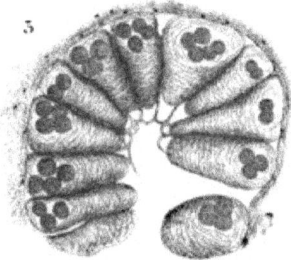

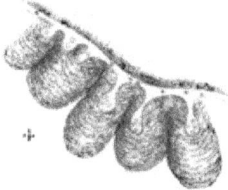

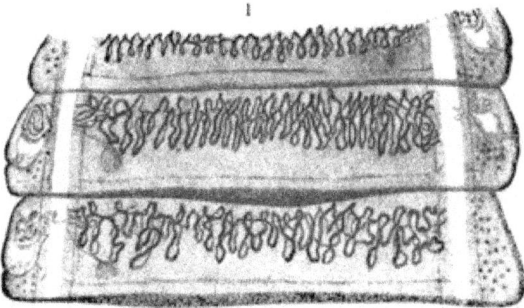

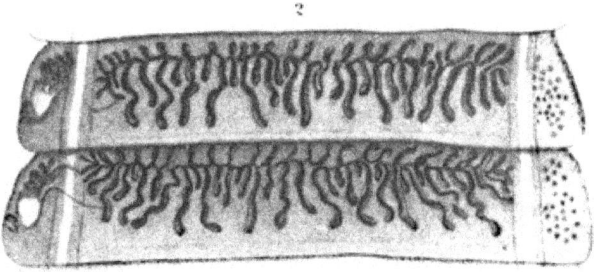

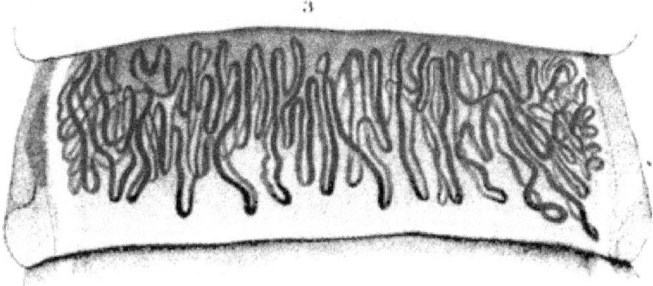

PLATE XIV.—*Stilesia globipunctata.*

Fig. 1. Strobilæ, fragments, natural size.

Fig. 2, 2b. Scolex, viewed ventrally, *en face* and from the side. ✕

Fig. 3. Segments 15^{mm} from head, showing nerves, canals and an organs; dorsal view. About 56 μ long by 0.57^{mm} broad.

Fig. 4. Segments 27-50^{mm} from head; the anlagen of the uteri are view. About 96 μ long by 1.14^{mm} broad.

Fig. 5. End of same, greatly enlarged.

Fig. 6. Right end of slightly older segments, showing ova in th dorsal view.

Figs. 7 and 8. Segments showing two stages of the uterus and tr fig. 7, ventral view; fig. 8, dorsal view; in fig. 7 the ovary is still about 0.16^{mm} by 1.4^{mm}; fig. 8, about 0.13^{mm} by 2.3^{mm}.

Fig. 9. End segments. About 0.26^{mm} by 0.64^{mm}.

Fig. 10. Egg. Zeiss ½ im.

130

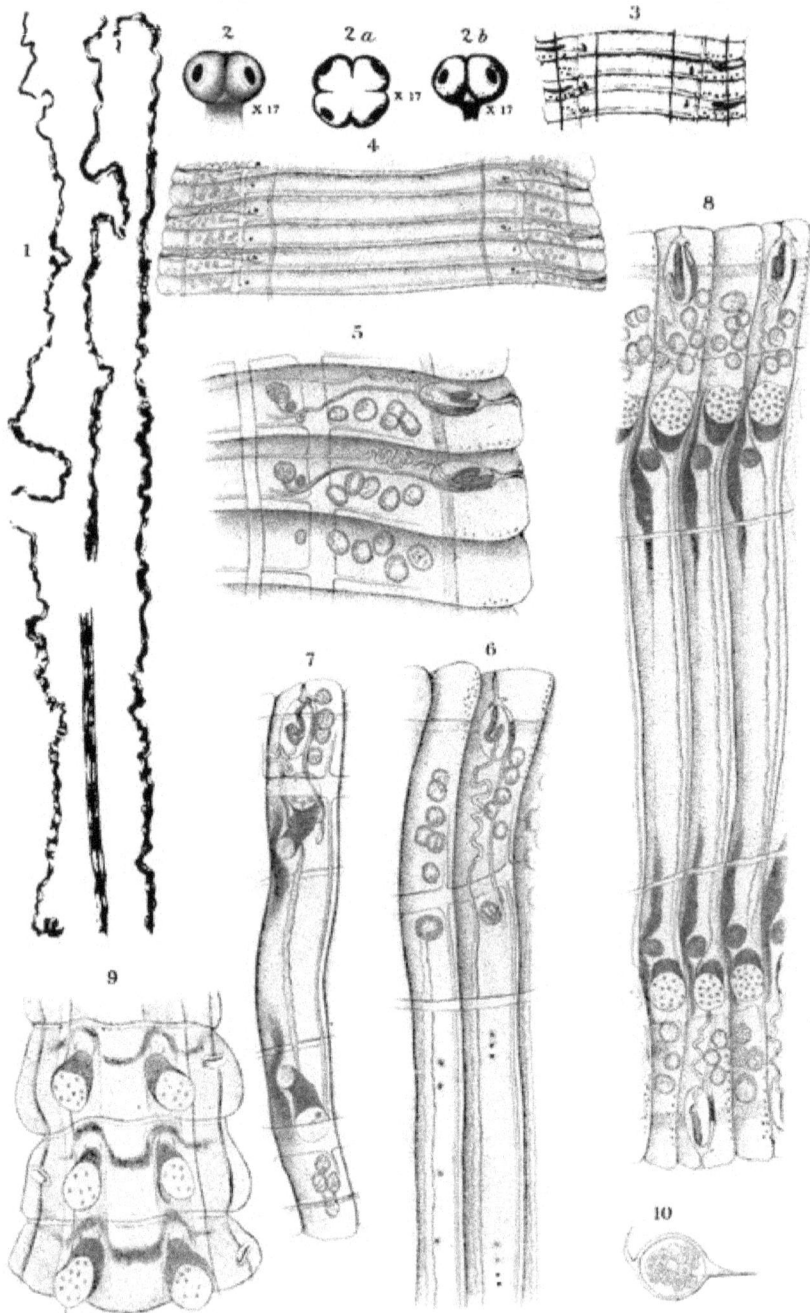

STILESIA GLOBIPUNOTATA

PLATE XV.—*Stilesia centripunctata.*

Fig. 1. Portion of a strobila, natural size.
Fig. 2. Head, ventral view. × 17.
Fig. 2a. Head, lateral view. × 17.
Fig. 2b. Head, *en face.* × 17.
Fig. 2c. Head, after Neumann.
Fig. 3. Segments 1^{mm} broad.
Fig. 4 Segments 1.12^{mm} broad.
Fig. 5. Segments 1.24^{mm} broad.
Fig. 6. Segments 1.6^{mm} broad.

132

PLATE XV

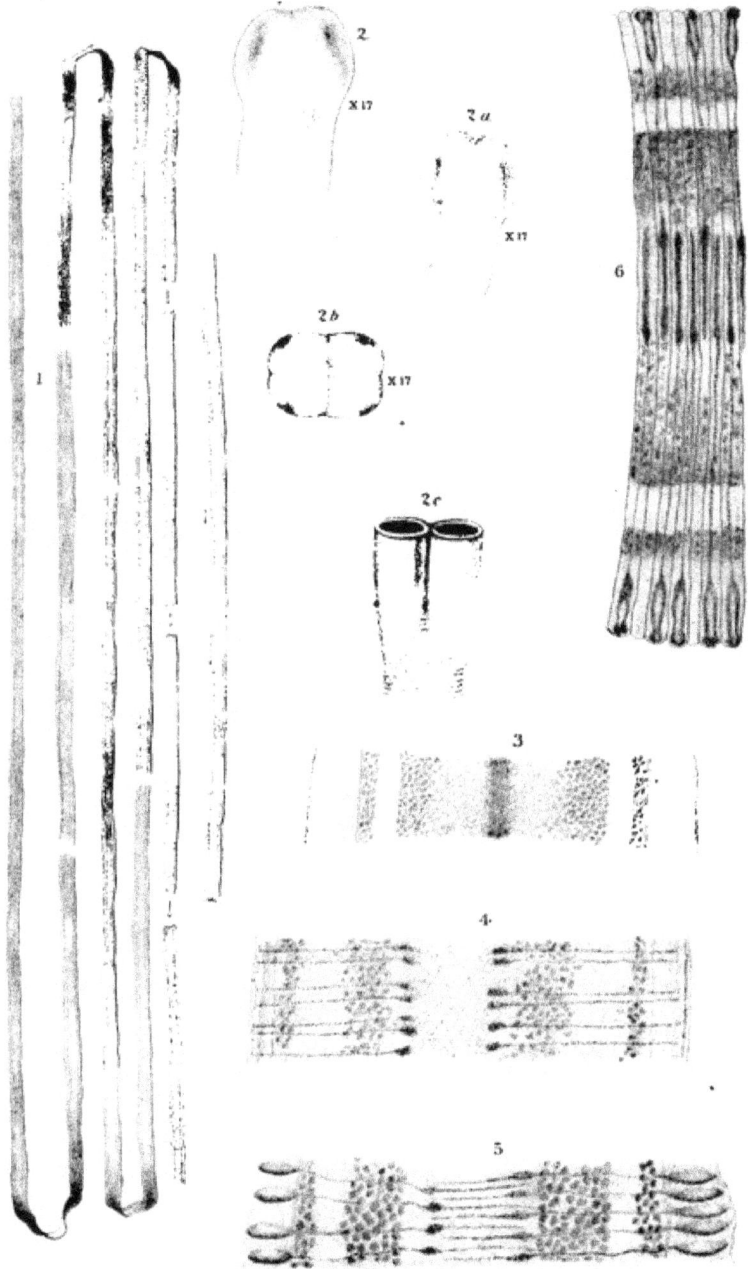

PLATE XVI.—*Moniezia, species inquirenda*.

Figs. 1–2*b*. Bureau collection, specimen No. 607, borrowed from Hassall's private collection, collected in England.

Fig. 1. Segments, natural size.

Fig. 1*a*. Segments. × 3.

Fig. 2. Head, viewed *en face*, alcohol specimen, much contorted. × 17.

Fig. 2*a*. The same, ventral view. × 17.

Fig. 2*b*. The same, lateral view. × 17.

Figs. 3, 3*a*. Head of Moniez's specimen of *M. nullicollis*.

Fig. 3. Head, lateral view. × 17.

Fig. 3*a*. The same, ventral view. × 17.

Fig. 3*b*. The same, *en face*. × 17.

Figs. 4, 4*b*. Bureau collection, No. 725, *Moniezia sp.*, Expansa group.

Fig. 4. Head, ventral view. × 17.

Fig. 4*a*. The same, lateral view. × 17.

Fig. 4*b*. The same, *en face*. × 17.

Figs. 5, 5*a*. Bureau collection, No. 612, *Moniezia sp.*, Planissima group.

Fig. 5. Head, ventral view. × 17.

Fig. 5*a*. The same, *en face*. × 17.

Fig. 6 and 6*a*. A head found in sheep. Possibly the same as B. A. I. collection No. 612.

134

PLATE XVI

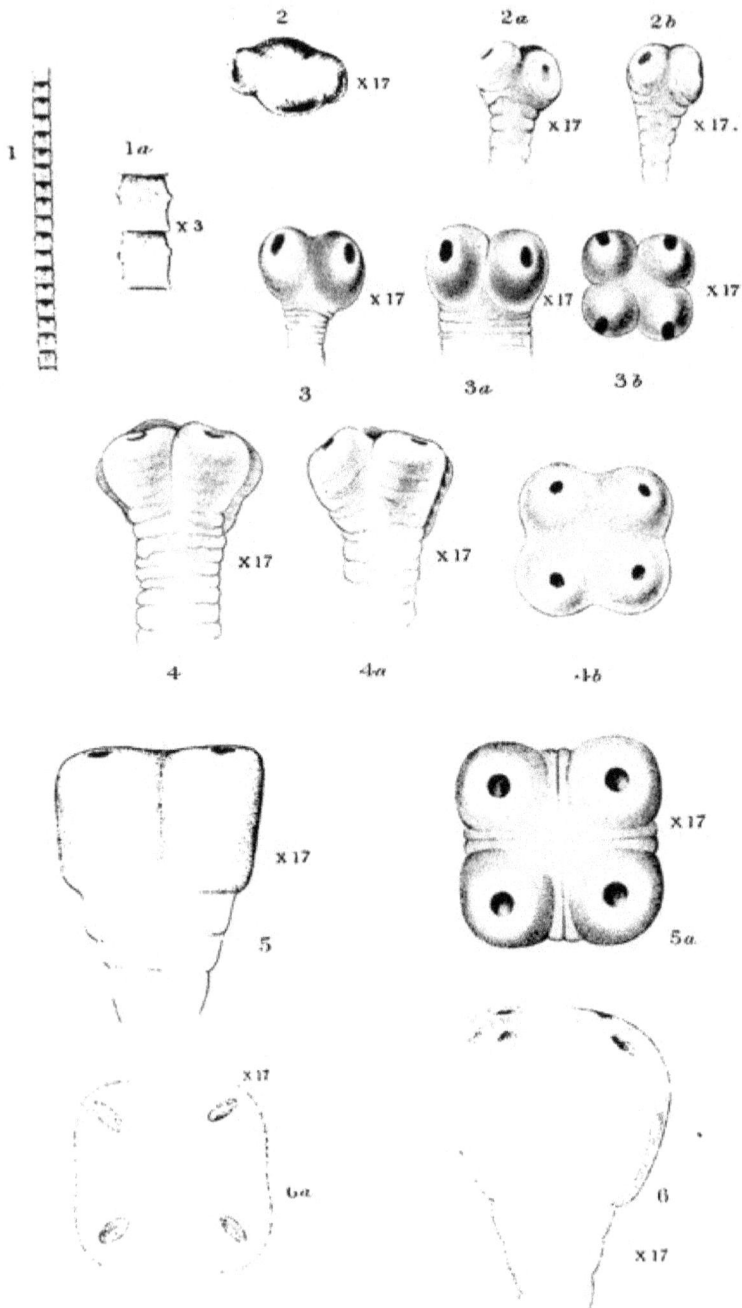

www.ingramcontent.com/pod-product-compliance
Lightning Source LLC
Chambersburg PA
CBHW021814190326
41518CB00007B/584